# Case Studies in Innovation

For Researchers, Teachers and Students

**Edited by**

**Heather Fulford**

Case Studies in Innovation
Volume One
First published: June 2012
Second Printing: February 2013

ISBN: 978-1-908272-37-9


Note to readers.
Some papers have been written by authors who use the American form of spelling and some use the British. These two different approaches have been left unchanged.

Published by: Academic Conferences and Publishing International Limited, Reading, RG4 9SJ, United Kingdom, info@academic-publishing.org

Printed by Lightning Source POD

Available from www.academic-bookshop.com

# Contents

# List of Contributors

***Claire Auplat***, *Advancia School of Entrepreneurship, Paris, France*
***Lynne Baxter***, *The York Management School, University of York, York, UK*
***Pere Caminal Magrans,*** *Technical University of Catalonia, Barcelona, Spain*
***Carine Deslee,*** *Université de Lille 2, France*
***Manel González-Piñero,*** *Technical University of Catalonia, Barcelona, Spain*
***Susanne Gretzinger,*** *University of Southern Denmark, Denmark*
***Saskia Harkema,*** *The Hague University of Applied Sciences, The Netherlands*
***Holger Hinz,*** *University of Flensburg, Germany*
***S Jacobs***, *Tshwane University of Technology, Pretoria, South Africa*
***Panayiotis Ketikidis,*** *CITY College – International Faculty of the University of Sheffield, Thessaloniki, Greece*
***Elena López Cano,*** *Technical University of Catalonia, Barcelona, Spain*
***Miguel Ángel Mañanas Villanueva,*** *Technical University of Catalonia, Barcelona, Spain*
***Wenzel Matiaske,*** *Helmut-Schmidt-University Hamburg, Germany*
***Arcot Desai Narasimhalu,*** *Singapore Management University, Singapore*
***Juan Ramos Castro,*** *Technical University of Catalonia, Barcelona, Spain*
***Cathie Wright,*** *School of Management and Languages, Heriot Watt University, Edinburgh, UK*
***Jonathan Youngleson,*** *Tshwane University of Technology, Pretoria, South Africa*
***Nikos Zaharis,*** *South East European Research Centre (SEERC), Thessaloniki, Greece*
***Sotiris Zigiaris,*** *Aristotle University of Thessaloniki, Greece*

# Introduction to Case Studies in Innovation

> *"In today's competitive landscape, the opportunities and threats happen swiftly and are relentless in their frequency, affecting virtually all parts of an organisation simultaneously. The business environment is filled with ambiguity and discontinuity, and the rules of the game are subject to constant revision. The job of management effectively becomes one of continual experimentation – experimenting with new structures, new reward systems, new technologies, new methods, new products, new markets, and much more. The quest remains the same: sustainable competitive advantage. Innovation and entrepreneurial actions represent the guiding light and motivating force for organisations as they attempt to find their way down this path."* (Kuratko, Goldsby and Hornsby 2012:4).

In *Leading Issues in Innovation*, the companion volume to this book, the editor, Daniele Chauvel, notes that "the concept of innovation has spanned R&D laboratories and organisation boundaries by extending to the whole organisation and its environment, integrating the duality of internal vs external sources of innovation" (Chauvel 2011:iv). This shift in focus, she comments further, has "a direct impact on actors" with the result that "in the knowledge society, any knowledge worker becomes a potential innovator".

The ten cases and research studies presented in this volume serve to illustrate the points that Chauvel has made regarding the reach and scope of the innovation function: in today's knowledge economy, innovation is about interaction and interfaces, collaborations and combinations, and enterprise and education. The studies have been selected from papers presented at the *European Conference on Innovation and Entrepreneurship* and published first in the refereed proceedings of that conference. They have been chosen because they present interesting and topical discussion points for both students and lecturers in the areas of innovation management, types of innovation, sources of innovation, knowledge transfer and exchange, entrepreneurial processes, and commercialisation. Each study is prefaced by some brief editorial commentary, together with some indications of specific topics that could be discussed in relation to the study. It is envisaged also that these discussion topics could act as useful points of departure for further research and academic inquiry in these areas.

The cases cover a variety of organisational types, including small businesses, medium-sized enterprises, large corporates, and public sector organisations. They incorporate discussion of small-scale R&D laboratories, larger higher educational institutes, through to industrial-sized research and commercialisation establishments. With regard to geography, the cases are drawn from a range of locations, providing an international focus to the topics and themes under discussion.

Despite the diversity that can be seen in the cases and research studies discussed in this volume, many of the challenges associated with successful innovation emerge as being remarkably similar. These include the identification, management and resolution of the tensions between, for example, creativity and commercialisation, education and enterprise, invention and innovation, collaboration and competition, know-how and knowledge exchange, resources and risks. A common thread running through these diverse studies is that when organisations of different types break down their traditional organisational boundaries, and when individuals of different types work together, then innovation and entrepreneurial processes can succeed, and true knowledge transfer can take place, leading to wealth creation and economic development. A key theme that emerges is the important role to be played by universities in the development and commercialisation of innovations, and in the transfer of knowledge to businesses in order to foster economic and regional development. The studies draw out discussion of topics such as the value of cross-disciplinary teamwork, leadership and strategic thinking in innovation management, networking, and partnering with organisations of different types.

Taken together, the studies provide an important reminder that innovation is not an isolationist activity: it is relational, and a key to its success is identifying the important relationships and ensuring that their quality is maintained and nurtured. So, this is a book of cases and research studies about innovation and entrepreneurial thinking, but underlying each study is not so much a traditional emphasis on innovation processes or the entrepreneurial process, but rather an emphasis on the people involved in innovation and on innovation partnerships.

> *"[...] innovative thinking is an integrated mind-set that permeates individuals and organisations in an effective manner"* (Kuratko, Goldsby and Hornsby 2012:4).

**Heather Fulford, 2012**
Centre for Entrepreneurship
Aberdeen Business School

**References**

Chauvel, D. (2011) *Leading issues in innovation research*. Volume One. Reading: Academic Publishing International Ltd.

Kuratko, D. F., Goldsby, M. G. and Hornsby, J. S. (2012) *Innovation acceleration: transforming organisational thinking*. New Jersey: Pearson Education, Inc.

# About the Editor

Dr Heather Fulford is the Reader in Entrepreneurship and Academic Director of the Centre for Entrepreneurship at Aberdeen Business School, Robert Gordon University, Scotland. Her research interests include entrepreneurship education and education resources, the language of entrepreneurs, social enterprise start-up and social entrepreneurship education.

Heather is currently supervising a number of doctoral students in aspects of entrepreneurship, entrepreneurship education and social entrepreneurship. She delivers courses at postgraduate and undergraduate level on new venture creation. Dr. Fulford is a visiting Fellow at Loughborough University, a Member of the British Computer Society, and a Fellow of the Chartered Institute of Linguists.

# From Technological Innovation to Innovation in Management Practices: The Case of Episkin®

**Claire Auplat**
Advancia School of Entrepreneurship, Paris, France

**Editorial commentary**

In this case study, Auplat traces the fascinating journey of a technological innovation (Episkin®) from its development at L'Oréal through to its influence on management practices in the organisation, and on to the point that it became a source of innovation in management practices. Furthermore, the case illustrates how this technological innovation helped pave a way for new cosmetics industry regulations to be implemented, and then led to the shaping of additional (and innovative) regulation in the industry.

This case study was constructed using archival data from the organisation, followed by a set of focussed interviews. This triangulated approach has been adopted to help ensure the robustness of the research findings, a factor which is particularly important given the original contribution the case makes to the wider innovation research arena through its examination of the relationship between technological innovation and innovation in management practices.

Some points for discussion and learning arising from this case include:

- The pros and cons of centralising the R&D function in an organisation.

- Comparing and contrasting the journey made by the technological innovation Episkin® with technological innovations in other organisations.
- The impact of a technological innovation on management practices.
- The contribution of a technological innovation to management strategy.
- The relationship between innovation and regulation.
- The role played by institutional entrepreneurs in shaping an organisation's innovation strategy.

---

**Abstract**: Innovation research has come a long way since the early models of linear innovation developed at the end of the 1940s, and it has diversified into many subfields. This paper is part of a stream which focuses on the relationships between technological innovation and innovation in management practices. It explores how technological innovation may be a source of innovation in management practices. It is based on the empirical case study of the development of Episkin®. Episkin is an in-vitro solution which is used by firms to test new products on engineered tissues. It allows scientists to screen and measure with high precision, the irritation, penetration, metabolism, or efficacy of a large number of formulations or active ingredients on reconstituted artificial skin and/or epidermis obtained by cell culture, in order to eliminate unsuitable ones. The development of this alternative method has led to a phasing out of animal testing by the cosmetics company L'Oréal and has accompanied regulatory change at the institutional level. Using a methodology which rests on the analysis of archival sources complemented by focused interviews, the paper firstly analyzes the process of technological innovation which led to the development of Episkin in the SkinEthic Laboratories, which are owned by L'Oréal. It then studies why and how this technological innovation was part of L'Oréal's management strategy. Finally, it shows that L'Oréal's use of technological innovation became a source of innovation in management practices. The paper aims to contribute to the important literature which is devoted to the study of various forms and sources of innovation. By its implications for policy making and management practices it also contributes to the strategic management literature, as well as to the policy literature.

**Keywords**: innovation, technological change, innovative management, Episkin, L'Oréal, REACH

# 1 Introduction and theoretical background

How can technological innovation become a source of innovation in management practices? This is what this paper investigates by looking at the particular case of the development of Episkin® and of practices concerning safety testing of cosmetics and other consumer products. Innovation is a fairly recent research domain and researchers started exploring the process of the implementation of invention roughly at the end of the Second World War. Early work focused mainly on the process of the diffusion of innovation, and the issue of understanding its sources attracted interest even more recently.

Rogers (1962) worked on establishing a typology of the adopters of a new innovation or idea. His studies showed that when the diffusion of innovation spread over various periods of time, the process could be modelled as an S-curve which reflected proportions of 2.5% innovators, 13.5% early adopters, 34% early majority, 34% late majority and 16% laggards. Alongside work on these categories and on the impact of the S-curve model on marketing, other scientists sought to understand the various gates that an invention had to go through before gaining the status of innovation. They built on the concept of a linear model of innovation which suggested that technical change happened in a linear fashion from invention to innovation to diffusion.

This linear model, which also originated at the end of the Second World War with no clear specific author (Bush 1946; Godin 2006), was criticized for being too static and for focusing too much on the role of scientists at the origin of inventions to the detriment of other players. However, it produced very important insights. In the early 1960s the industrial innovation process was usually perceived as a linear progression from scientific discovery through technological development in firms to the market place. This was the 'technology push' model, popularized by Rothwell (1994). This model accounted for a period of intense technological activity coupled with the necessity to rebuild the country after the war. It was pitted against the 'Market pull' model of the early 1970s (Myers & Marquis 1969), which considered that the production of innovation resulted from market demand: there was a business need for a product, and technology developed to fill that need.

Kamien & Schwartz (1975) did a complete literature review of the work done on the connections between innovation and market structure, and they found that the various analyses they surveyed led in the end to a lack of robust findings concerning notably the importance of firm size, monopoly power and inventive activity. Then, Dodgson, Gann, & Salter (2005) produced a typology of the range of research in innovation. They identified five major focuses of innovation research: the source of innovation, the nature and extent, the type, the system, the process and the outcomes.

Building on all this previous work, this study explores one additional dimension: the connection between technological innovation and innovation in management practices. Its specific objective is to examine how technological innovation may be a source of innovation in management practices. The research question which guides the analysis is 'How did the development of the technological innovation of Episkin® lead to a new approach of ingredients safety testing in cosmetics and other products?

## 2 Methods

The paper is based on the empirical case study of the development of Episkin®. Episkin produces standardised samples of reconstructed skin to evaluate tolerance and effectiveness of ingredients in the chemical, pharmaceutical and cosmetic industries. This testing model allows scientists to screen and measure with high precision the effects (irritation, penetration, metabolism, or efficacy) of a large number of formulations or active ingredients in order to eliminate unsuitable components. The development of this alternative method led to a radical change in management practices: the replacement of animal testing by synthetic skin testing to check the safety of cosmetics and other products.

This case study (Yin 2003) uses a qualitative method based on archival work completed by focused interviews (Glaser & Strauss 1965). Using focused interviews to complete archival work is part of the triangulation method advised by Jick (1979) to increase the robustness of findings. The paper firstly analyzes the process of technological innovation which led to the development of Episkin in the SkinEthic Laboratories, owned by L'Oréal. The second part outlines the development of cosmetics regulation leading to the banning of animal testing. The paper then studies why and how the technological innovation of Episkin became part of L'Oréal's management strategy. Finally, it shows that L'Oréal's use of technological innovation became a source of innovation in management practices.

## 3 Episkin, SkinEthic and technological innovation

Innovation has been a constant trait at L'Oréal since the origins of the company in 1908 (Marseille 2009). EugèneSchueller, its founder, was a chemist. When Charles Zviak became L'Oréal's third CEO in 1984, the company was one of the very few large multinational companies to be headed by a chemist. By 1950, it had a research-and-development team of 100 chemists and the number increased to 300 in 1960, to reach 750 in 1974. In 1920, L'Oréal had already three research chemists, which was a large number considering that the firm had only ten sales representatives. In 2010, the R&D sector employed 3,313 employees of 60 different nationalities who worked in 18 research centres and 13 evaluation centres globally. Innovation was and remained at the heart of the company's development. L'Oréal filed 674 patents in 2009. In terms of management, R&D was the only area which remained centralized apart from overall management control. Indeed, the group consisted of a federation of competitive companies including 147 production and distribution facilities worldwide and so all the other business activities were decentralized. Besides, since 2003 and in spite of the economic crisis, the group still spent an average of about 3.4 percent of its revenue on R&D, which was nearly one percent more than its major competitors Unilever and Procter&Gamble.

Since the beginning of its operations, the company showed flair in understanding the importance of pursuing innovation in a sustainable way (Auplat & Lapoule 2011). From the start, L'Oréal's top management had the vision that innovation was destined to increase the beautification of customers, but in a way that was both pleasant and safe. What is of particular interest for this study is the innovation process which led to the 1997 landmark development of Episkin. L'Oréal started working on reconstructed skin in its research facilities as early as 1979. One of its researchers, Marcelle Régnier, succeeded in carrying out the first reconstruction of human epidermis. The breakthrough happened in 1983 at L'Oréal's research laboratory in Aulnay in France. Roughly at the same time, another French researcher, Estelle Tinois, had developed a kit of artificial skin containing collagens. She worked at Imedex, a research laboratory in Lyon, France, which had also been working on collagens used for reconstructive surgery. Imedex belonged to the Mérieux Institute, itself part of the Rhône-Poulenc Group. There, Estelle Tinois developed a skin testing method including cell culture testing and testing on reconstructed skin models to evaluate tolerance and effectiveness of ingredients and finished products. Her kit of artificial skin was called Episkin. L'Oréal started a partnership with Imedex in 1994, bought the Episkin process and research team in 1997 and then set up an Episkin subsidiary in Lyon-Gerlan in 2001. L'Oréal invested a total of sixteen million Euros in the project over a decade, and inaugurated in April 2011 the L'Oréal Predictive Evaluation Center, the first Global Predictive Evaluation Centre for the cosmetic industry in the world.

When L'Oréal developed the Episkin Company in 2001, it had only two major competitors working on reconstructed skin, Matexk in the United States, and a company located in Nice, France, called SkinEthic(Le Progrès de Lyon 2000; 2011). SkinEthic, a company founded in 1992, had developed in the early 90s a cell culture process allowing to mass-produce in vitro various models of reconstructed human tissues with specific characteristics. It soon became the world leader in tissue engineering for in-vitro human skin models and developed an expertise in producing robust reproducible models to allow the screening of a large number of raw materials, active ingredients and finished products in various forms (solid, liquid, powder, cream or gel). It was listed on the Marché Libre of Euronext Paris in 2004. However, in spite of a forecast of € 1.8 million annual sales and 0.83 million net profit for 2004, SkinEthic had annual sales of €1.47 million and a net loss of € 0.31 million for that same year.

Episkin bought 81.2% of the share capital of SkinEthic from the company's founder Martin Rosdy in 2006, as well as 2.9% of the share capital owned by employees. The acquisition, which combined the expertise of Episkin and of SkinEthic, allowed L'Oréal to create a top-level industrial facility capable of meeting the needs of the entire industry. In spring 2011, the Episkin facilities at Lyon could routinely produce nine models of skin and cornea, and some 130,000 units of reconstructed skin tissues per year, against 16,000 per year in 2000. They could test corrosiveness, skin or eye irritation, allowing safety tests on 1000 products and efficacy tests on 100 products. Most importantly, the Episkin test method was approved by the European Chemicals Agency (ECHA) and by the Organization for Economic Co-Operation and Development (OECD) as an alternative to animal tests.

# 4 Institutional innovation

There were examples of laws aiming to protect consumers from unsafe foods and medicines as early as 1202 with King John's English food law. However, food and health regulation really started in the US at the turn of the 20th century. The Biologics Control Act was passed in 1902 to ensure purity and safety of serums, vaccines, and similar products used to prevent or treat diseases in humans. At that time, the US Congress commissioned a study of chemical preservatives and colours and their effects on digestion and health. In the wake of public outcry after the publication of a novel describing the appalling conditions in which meat was handled in Chicago stockyards, followed by further disclosures of the use of poisonous preservatives and dyes in foods and dubious medicines, the US Congress passed the Pure Food and Drug Act in 1906, and then the Sherly Amendment and the Meat Inspection Act. These regulatory documents constituted a legal framework which was destined to guarantee the quality of the food and medicines used by the consumers. Other landmarks in consumer protection were the establishment of the Food and Drug Administration (FDA) in 1931, and of the Food, Drug and Cosmetic Act of 1938 which enabled the FDA to test new drugs before they reached the hands of the consumers. The Consumer Product Safety Commission (CPSC) was created in 1973 to guarantee the safety of more than 10,000 consumer products in the US.

In the EU, REACH (the Registration, Evaluation, Authorization and Restriction of Chemical Substances) was introduced by the European Commission as a comprehensive regulatory framework in 2006 to replace some 40 existing legal acts and to create a single system for all chemical substances in Europe (European Union 2006). Its main objective was to improve the protection of human health and the environment through the better and earlier identification of the intrinsic properties of chemical substances, and this included the plan to find substitutes for the most dangerous chemicals currently in use. The law entered into force on 1 June 2007. One of its characteristics was that it transferred the burden of proof for demonstrating the safe use of chemicals from EU Member States to industry, and it is now the responsibility of manufacturers and importers to produce the safety assessment of their products.

REACH had two major components. Starting in June 2008 and before the deadline of June 2018, all firms that manufactured or imported more than one tonne of a chemical substance per year were required to register it in a central database – the European Chemical Agency (ECHA). Besides, all new chemical substances that were produced or imported in the EU after the cut off date of 1981 had to undergo a strict risk assessment process. The REACH provisions were planned to be phased in gradually over a period of 11 years because of the complexity of the process and because of the need to find replacements for some chemicals that were considered too dangerous to be retained.

One of the effects of REACH was to increase the number of tests to be carried out, which entailed increasing animal testing. The history of testing for consumer products safety goes back in time several thousand years. Already in the time of ancient Egypt people used the services of food testers to verify that their food had not been poisoned. At that time, safety testing relied on testing products on human slaves and on animals. Systematic testing for product safety on a large scale also took off after the Second World War in the 1950s with the rise of modern epidemiology. Until very recently, safety tests were done on animals, and the progress of science and regulation led to a paradox: on the one hand there was a demand for a ban on animal testing, but on the other hand, the new evaluation programs implemented by the EU and other regulatory bodies led to the testing of an increasing number of products. For example, in 2009 the EU counted two thousand cosmetics companies. Out of their € 60 billion turnover, about a quarter came from new products, some 5,000 of them per year. The new EU regulations meant that all these products had to undergo safety assessments and this involved tens of thousands of safety tests to cover all the new ingredients.

# 5 Discussion and conclusions

The study of the development of Episkin shows the importance of L'Oréal's pro-activity in the phasing out of animal testing, and therefore on the adoption of new management practices. Pro-activity happened on three levels.

- R&D support and funding. L'Oréal's researchers working on reconstructed skin were among the first ones globally. The company's strategy was to invest heavily in infrastructures and to buy additional competencies to become a world leader in synthetic skin testing. It developed and validated testing methods of safety assessment which combined tissue engineering with a range of predictive methods that included statistical and mathematical models, via computer simulation. The entire raw materials catalogue of L'Oréal was tested using these models and a total of over 12,000 formulas and 2000 ingredients were evaluated since 2006. New ingredients were systematically being tested using these models. By doing this, the company developed a new benchmark of safety testing. This was done on a voluntary basis, and it involved heavy investment. However it was a shrewd business move in view of the upcoming trend in regulation concerning animal testing.

- Communication. L'Oréal communicated regularly on the fact that the adoption of Episkin and in vitro testing led to two very important developments. On the one hand it meant that the group had stopped using animal testing for the evaluation of its entire range of finished cosmetic products in 1989, and that it could gradually implement systematic testing of all the ingredients it used so that in 2010, 100% of the ingredients it used were tested. On the other hand, L'Oréal reminded consumers and the general public that innovation was a way to help reduce the number of animal tests carried out and that in Europe, there were around 30,000 animal tests a year in the early 2000s. This was important for animal rights campaigners, but also for consumers. There was indeed growing uneasiness with the idea that the introduction of new products on the market went through animal testing and this was especially true for cosmetics, which were usually non-essential consumer products.
- Institutional work. At the same time as it developed its technological knowledge, L'Oréal took part in several workshops organised by regulatory agencies. It was involved in a number of international programmes of research related to or conducted to regulatory design. These included the International Council on Nanotechnology (ICN), the Center for Biological & Environmental Nanotechnology (CBEN), the World Business Council for Sustainable Development (WBCSD), the European Union Scientific Committee on Consumer Products, and the EU Working Group on Nanosubstances in Cosmetics. The group also worked with several working groups of the In-

ternational Organization for Standardization (ISO), notably the working groups on ISO nanostandardisation, and groups working on the elaboration of the ISO 26000 norm, a norm on social responsibility.

By developing new technologies and actively promoting them with the public and with regulatory agencies, L'Oréal was able to become an institutional entrepreneur (Garud, Hardy, & Maguire 2004; Lounsbury 2007): it contributed to changing existing institutions so that the safety of consumer products could be assessed by means other than animal testing. The EU DIRECTIVE 86/609/EEC on the use of experimental animals stipulated that experiments should not entail the use of an animal if another scientifically satisfactory method could be used in a practical way (European Union 1986). The 7th amendment of the European Cosmetics Directive set a progressive ban starting in 2009 on the use of animal testing in Europe for the assessment of cosmetic ingredients (European Union 2003). However, at the same time, the only alternative methods which were systematically adopted at Community level were those which had been scientifically validated by the European Centre for the Validation of Alternative Methods (ECVAM) or the Organisation for Economic Cooperation and Development (OECD 2006; Spielmann 2009). As ECVAM and OECD both validated the Episkin method, this was the key leading from technological innovation to innovation in management practices.

Regulation L220 of 24 August 2009 of the Official Journal of the European Union, a regulation which is binding in its entirety and directly applicable in all Member States of the EU, amended previous regulations concerning the safety testing of ingredients "for the purpose of its adaptation to technical progress". The amendment explained how animal testing should be replaced by synthetic skin testing and how the Episkin model had been validated as an alternative to animal testing (Official Journal of the European Union 2009; Spielmann et al. 2007).

This case illustrates the path that went from technological innovation to innovation in management practices. The EU DIRECTIVE 86/609/EEC on the use of experimental animals outlined clearly the objective of reducing animal testing for the safety tests of new ingredients as early as 1986. However, it stipulated that this could only happen when other scientifically satisfactory methods could be used in a practical way. It was not in a position to provide the solution, it only stated the objectives. It was because the L'Oréal Group had both the vision and the financial power to develop the Episkin model that there could be a shift from animal testing to artificial skin testing. By promoting technological innovation, it created the benchmark that could be used for innovation in management practices: it was in a position to test all the ingredients of the raw materials it used before this was made obligatory by international regulation and this proactivity allowed institutions like ECVAM and OECD to verify that the alternative solutions to animal testing it developed worked, and could be validated at community level.

## References

Auplat, C. and Lapoule, P. (2011) "L'Oréal: science, innovation... and sustainable development", CCMP, Case and Teaching note.

Bush, V. (1946) Science: the endless frontier, Washington: National Science Foundation, reprint 1960.

Dodgson, M., Gann, D., and Salter, A. (2005) Think, Play, Do - Technology, Innovation and Organization, Oxford University Press.

European Union (2003), "7th amendment to the European Cosmetics Directive DIRECTIVE 2003/15/EC OF THE EUROPEAN PARLIAMENT AND OF THE COUNCIL of 27 February 2003 amending Council Directive 76/768/EEC on the approximation of the laws of the Member States relating to cosmetic products".

European Union. (1986) "COUNCIL DIRECTIVE of 24 November 1986 on the approximation of laws, regulations and administrative provisions of the Member States regarding the protection of animals used for experimental and other scientific purposes (86/609/EEC)".

European Union. (2006) "Regulation (EC) No 1907/2006 of the European Parliament and of the Council of 18 December 2006 concerning the Registration, Evaluation, Authorisation and Restriction of Chemicals (REACH)".

Garud, R., Hardy, C. and Maguire, S. (2004) "Special Issue on 'Institutional Entrepreneurship'", Organization Studies, 25(8): 1471.

Glaser, B. G. and Strauss, A. L. (1965) "Discovery of substantive theory: a basic strategy underlying qualitative research", American Behavioral Scientist, 8: 5-12.

Godin, B. (2006) "The Linear Model of Innovation", Science, Technology & Human Values, 31(6): 639-667.

Jick, T. D. (1979) "Mixing Qualitative and Quantitative Methods: Triangulation in Action", Administrative Science Quarterly, 24(4): 602-611.

Kamien, M. I., & Schwartz, N. L. (1975) "Market Structure and Innovation: A Survey". Journal of Economic Literature, 13(1): 1.

Le Progrès de Lyon. (2000) "Lyon, capitale mondiale des tests de cosmétique", Le Progrès: 7. Lyon.

Le Progrès de Lyon. (2011) "De la création d'une crème virtuelle au test d'irritation sur la peau artificielle, Le Progrès: 6. Lyon.

Lounsbury, M. (2007) "New Practice Creation: an Institutional Perspective on Innovation", Organization Studies, 28(07): 993-1012.

Marseille, J. (2009) L'Oréal 1909 -2009, Librairie académique Perrin.

Myers, S. and Marquis, D. G. (1969) Successful Industrial Innovation, Washington: National Science Foundation.

OECD. (2006) "Test Guideline 435. OECD Guideline for the Testing of Chemicals. In Vitro Membrane Barrier Test Method", Available at: http://www.oecd.org/document/22/0,2340,en_2649_34377_1916054_1_1_1_1,00.html.

Official Journal of the European Union. (2009) "Legislation L 222". Available at: http://eur-lex.europa.eu/LexUriServ/LexUriServ.do?uri=OJ:L:2009:220:FULL:EN:PDF

Rogers, E. M. (1962) Diffusion of Innovation, New York: Free Press.

Rothwell, R. (1994) "Towards the Fifth-generation Innovation Process", International Marketing Review, 11(1): 7.

Spielmann, H. (2009) "20 Years of ZEBET 1989-2009: A Success Story", Paper presented at the 20th Anniversary of ZEBET at BfR and 50 Years of the 3Rs Principle, Berlin, Germany.

Spielmann, H., Hoffmann, S., Liebsch, M., Botham, P., Fentem, J., Eskes, C., Roguet, R., Cotovió, J., Cole, T., Worth, A., Heylings, J., Jones, P., Robles, C., Kandárová, H., Gamer, A., Remmele, M., Curren, R., Raabe, H., Cockshott, A., Gerner, I., and Zuang, V. (2007) "The ECVAM International Validation Study on In Vitro Tests for Acute Skin Irritation: Report on the Validity of the EPISKIN and EpiDerm Assays and on the Skin Integrity Function Test", ATLA, 35: 559-601.

Yin, R. K. (2003). Case study research. Design and methods, 3rd edition, California: Sage.

# A Case Study of Entrepreneuring: Redesigning Technologies for a Commercially Viable Cancer Detection Product

**Lynne Baxter[1] and Cathie Wright[2]**

[1]The York Management School, University of York, York, UK
[2]School of Management and Languages, Heriot Watt University, Edinburgh, UK

**Editorial commentary**

Baxter and Wright examine the processes of entrepreneuring using a case study involving the development of a cancer detection and treatment technology. They demonstrate that entrepreneuring takes place within a worknet of actors and interactions, rather than adhering to a strictly linear model. The case also highlights some of the tensions that can arise in the process of attempting to commercialise academic research. This case was developed through interviews with key stakeholders in the entrepreneuring process, together with analysis of relevant background documentation. Discussion of the case's implications is grounded in relevant theory of entrepreneurial process.

Some points for discussion arising from this case include:

- The role of knowledge relationships and the factors that lead to their success.
- The role of the CEO in a worknet.
- The value of worknets in the entrepreneurial process.
- The contribution that academics can make to the commercialisation process.

**Abstract**: There is considerable appeal in defining a linear process of entrepreneuring, mirroring the activities required to successfully launch a new business or product. However various authors have noted that business life is not quite so simple. This paper explores the various processes of entrepreneuring (MacMillan 1986, Steyaert and Katz, 2004) and examines them in relation to a case study of the commercialisation of a new form of technology that detects and visually represents cancerous and precancerous cells. The technology is proven to work, and in 2009 a new CEO was appointed to drive the commercialisation; key to this is the creation and development of a relationship with the subcontractor design and manufacturing company. This relationship resulted in a completely new ergonomic design that more effectively met the requirements of the clinical process. It also had the benefits of being modular, enabling segmentation of the product for different markets; greatly reducing the cost of production, at the same time as facilitating its ability to be manufactured. The role of knowledge relationships and their importance in the mobilisation process are discussed in terms of relating this to the concepts of Actor Network Theory (ANT) (Callon 1986) and Interaction Rituals (Collins 2004). The paper demonstrates how complex and shifting these interaction relations are, but essential to the entrepreneuring process. We argue that the CEO is an important actor, but not the only one important in the success of the business. The authors propose that entrepreneuring takes place in a "worknet" (Latour 2010) of many people performing simultaneously many strands of activities, in no way a linear discrete process overall, but one which contains short sequences of linearity.

**Keywords**: entrepreneuring, product development, power, knowledge

# 1 Introduction

Over a nine year period an academic investigated methods of detecting and visually representing cancerous and precancerous cells of the cervix, and invented a new form of technology. A university spin-out company was formed and secured initial funding to commercialise this invention, but after several years this vehicle was deemed unsuccessful. A new CEO was appointed as part of a process of reorganization and the original inventor was marginalised. The product had contained proven advances in detecting cancer but was neither ergonomic to use nor economic to produce. It could only be produced on a made to order unit basis and was ill-suited to the clinical environment.

The paper will explore various processes of entrepreneuring (MacMillan, 1986; Steyaert and Katz 2004) effectuated by the new CEO (Sarasvathy, 2001) to date, who has arrived at a point where the technology is proven to work, but the form has had limited appeal. Successful commercialisation depends on his performing several roles simultaneously in relationships with several important others. All these things had to be effectuated at a time when there was a North Atlantic recession and the money supply was tight.

In line with Callon (1986), the CEO's ability at entrepreneuring rested on his knowledge bases and skills at mobilising key stakeholders in the technology development process in a way that the original inventor could not. The paper is based on qualitative interviews with key stakeholders and company documents. We argue that in order to have skills at entrepreneuring you do not necessarily have to have been a stereotype entrepreneur from birth, that entrepreneuring is born out of the situation and chains of interactions through which the people effectuate innovation. The paper looks at the role of knowledge relationships and their importance in the mobilisation process, showing how complex and shifting interaction relations are, but essential to the entrepreneuring process.

The paper is divided into three sections. The first encompasses theories of entrepreneuring in order to locate our argument in relation to previous work. The second is where we describe the technology development, focussing on the actors and interactions. This is to form a basis of understanding for the third and final part of the paper, where we relate our research on the development with the theories outlined in the first section and draw some conclusions about our work for the conference audience. Briefly, the technology development process has been lengthy, with a long period of relative stagnation after the initial invention, but after some battles and the establishment of new relationships, considerable progress has been made in the last two years.

## 2 Entrepreneurs, entrepreneurship and processes of entrepreneuring

There is no agreement on a definition of what constitutes an entrepreneur if some shared understanding about what they do, but recently there has been work which returns to exploring actual business and utilises social theory, where we would locate our work. A very disparate set of authors write on the topic - from Management, Sociology, Psychology, Economics and beyond - and they have not settled on whether entrepreneuring is about either a person or a process. The activities which make up the practice of entrepreneuring encompass for example invention, company formation, market creation, and these are performed by entrepreneurs which some authors think are a class and therefore numerous and others think are special and have to be nurtured. Our preference is to see entrepreneuring as an assemblage (Deleuze and Guattari, 1987) of both intersecting and discontinuous processes performed in situated interactions (Collins, 2004; Goss, 2005) by several and parts of individual actors which can be human or non-human (Latour, 2010). The different actors will have their own view of the processes, based on their past experience and knowledge bases, ie there is no single view of the 'network' or what the technology is, and views on its development are polyphonous. We think it is important to note the processes of entrepreneuring will have hiatuses and discontinuities and recursive elements, and agree with Schumpeter's insight about innovation being a 'creative disruption' – not all activities will be positive or functional.

Van Praag (1999) provides a review of the early thinking, including Cantillon, who saw entrepreneurs as fulfilling a useful arbitrage function in bringing two classes possessing capital and labour together for a profit, which had its risks for the entrepreneur. The basic cornerstone of 'arbitrage' was retained in Say's later discussion retold in Van Praag (1999), which added a leadership role within their own organization. Performing these activities simultaneously took knowledge of the specific market as well as experience, which few people possessed and so the role commanded good remuneration. These two definitions in the main refer to the activities of being an entrepreneur with the knowledge bases of the person included but not the centre of the definition. Entrepreneurs as the drivers of innovation emerge in the work of Marshall (1890, 1930), who retained the ideas from Cantillon and Say and added a discussion of their role as being on the look-out for opportunities to innovate in the market for their company. Entrepreneurs can be motivated by the desire to have a private kingdom (Van Praag, 1999: p321), an aspect noted by Schumpeter (1934), whose definition of entrepreneurship rests on the activities carried out, coming as it does from an economics perspective which Thornton (1999) describes more fully.

In her work, Thornton (1999, p20) defines 'entrepreneurship as the creation of new organizations (Gartner, 1988), which occurs as a context-dependent, social and economic process (Reynolds, 1991, Low and Abramson, 1997)', which we might see as a backdrop to the activities in previous definitions. She classifies writing on entrepreneurship as being either on the 'supply side' where the attributes of individuals who could perform the role are discussed, or 'demand side' where the focus is upon how markets and firms create new organizations, and has a very economics flavoured perspective on what is intended to be a sociology of entrepreneurship. Despite this, she alerts us to the importance of context in entrepreneuring, stating:

'While much has been learned about the personal attributes, behaviours and other characteristics associated with entrepreneurs there has been little progress in relating types of entrepreneurs to the formation of new ventures. Because of the challenges of founding new organizations vary by context, different types of enterprises are likely to require different types of entrepreneurs (Thornton 1999, p 23).'

McMullen and Shepherd (2006, p 132) define entrepreneurship as: 'To be an entrepreneur, therefore, is to act on the possibility that one has identified an opportunity worth pursuing.' They produce a model which is intended to bridge the gap between system level economics thinking on the topic, and individual more psychological or functionalist sociological perspectives. They stress the importance of trying to understand the concept of uncertainty more fully; their argument is structured to 'highlight the importance of judgement under uncertainty, and then acknowledge the role of belief and doubt in entrepreneurial judgement using a discussion involving three types of uncertainty: state, effect, and response (McMullen and Shepherd, 2006, p133).'

Howarth, Tempest and Coupland (2005) adopt a more European, Organization Theory perspective by employing Burrell and Morgan's four part classification (1979) modified by Alvesson and Deetz (2000) to discuss prior definitions of the entrepreneur, which they say are mainly in one quadrant, the 'normative' paradigm, couched in terms of a positivist ontology leaving three others to explore. Howarth et al (2005) use the paradigms in their analysis of four cases of actual entrepreneurs to set up interplay between the ideas of the various authors in order to foreground the usefulness of the other paradigms.

Steyaert (2007, p 453), provides an example of other paradigm work outlining process approaches to entrepreneuring, noting that previous authors have tried to find a 'key concept that could elucidate the inherently process-oriented character of entrepreneurship'. Whether this is possible or not is unlikely ever to be resolved, but he charts the development of work in entrepreneurship away from Thornton's two category view towards more engagement with social theory and philosophical ideas, through research into relations and relatedness. An example of this is Fletcher (2006), who concludes (p 437): 'The activities we label entrepreneurship require immense effort, negotiation and dialogue and they always 'go on' in relation to something else that has gone before.'

This point links to one of Steyaert's process theories, Actor Network Theory (ANT), which stemmed from Callon (1986) attempting to discuss the power relations surrounding the development of technologies. ANT describes processes as consisting of four 'moments' of translation, and treats the technology as an actor in its own right. Recently Latour (2010) stated that technologies are always in the process of being 'redesigned' – there is nothing static or really new, and people have viewed the theory simplistically. He would rather the subject was about 'worknets' of associations made up of interwoven strands like Deleuze and Guattari's assemblage or rhizome. Earlier (1986), he contrasts ANT as a means of studying power relations with more orthodox class-based theories which explore the initiators, their activities and forms of resistance. In ANT, every step in the development of technology is a separate act of a person, and it is only in very peculiar circumstances that a smooth, obstacle-free process occurs. There is no initial impetus to create inertia where a discrete technology reaches the market unchanged; each person has to add some energy to the process. Latour uses the analogy of a rugby game. The energy that propels the technology cannot be hoarded, and for it to progress, individuals have to take an active role (co-ordinate, run and pass the ball), and this frequently changes the nature of the technology as it is continuously transformed.

If one subscribes to ANT, a smooth development process is highly unlikely. A key decider on whether a technology becomes popularised is if the people involved in creating it are skilled at forming coalitions of like-minded to guard and nurture the development forward. The 'sociology of translation' has 'moments' as follows: 1) problematisation where the 'enrolling agency' tries to convince others that they have the solution to their problem. They have to lead the target groups through a series of practices the original group can define and control. These are known as 'obligatory passage points'. 2) 'interessement' where relationships are fostered; 3) 'enrolment' occurs when the alliances and coalitions agree on what they want the outcome of the process to be; and 4) mobilisation, where the enrolling agencies control the enrolled to preserve the pattern of interests which is satisfactory to the enrolling agency.

Callon and Latour do not expect the process to be successful every time. The usual illustration quoted is Callon's (1986) scallop fishing off Brittany, where scientists were able to create a reputation and enrol interested parties such as government bodies, fishermen and, apparently, scallops into agreeing to participate in the scientist's preferred solution. The scallops and the propagating technology were treated as actors in the process with similar status to the others, and it was their 'action' or lack of it which contributed to in this instance the failure of the translation process. We think that entrepreneuring can be seen as a process of translation which takes place in a complex 'worknet' which includes technologies as actors which are continually being redesigned.

Latour (1993) argues that functionalist theories wrongly assume a unity and consistency to a person, who is in fact a network in him or herself - 'an actor is nothing but a network, except that a network is nothing but actors' (Latour, 2010, p 5). We should focus not on 'surfaces' but filaments between and disconnected from nodes, not the nodes themselves, which are points on many dimensions with many connections. Instead of trying to establish the 'most important' connection, we should investigate the range as several weak filaments form a strong and productive association. Like Steyaert (2007), he notes the importance of reversing the foreground/background activities and the flow of desire and affects. Latour (2010) exhorts us to set aside our preconceived ideas about distance and closeness when describing networks, as it is possible for close connections to be not helpful or constructive in the process and further away ones more so. Networks should not be evaluated on their scale either, more or fewer connections may not be important to the success or traditional concepts of hierarchy. A key feature of an actor network is that there is no 'outside' or fixed boundary for the network.

The day-to-day interactions of actors are key to this process and Goss (2005) draws on Collins (2004) to discuss how these might be theorized. He believes that social interactions often take a ritualized form which includes and excludes individuals – 'membership patterns and the ideas that accompany them ....ephemeral, stratified and conflictual' (Collins, 2004, p xi). Rituals foster or destroy emotional energy which 'gives the ability to act with initiative and resolve, to set the direction of social situations rather than be dominated by others in the micro-details of interaction' (2004, p 134).

We think that many of the steps in the technology development process involve interactions which take the form of rituals – for example meetings with customers and subcontractors, board meetings, presentations etc. He notes that many of these rituals ideally share common ingredients: they are embodied – people physically meet, they are private, the people concerned are interested in developing the technology and there is a shared mood. If these ingredients exist there is likely to be a positive emotion or as Collins (2004) puts it, 'collective effervescence', which gives momentum and solidarity to the network because the actors derive emotional energy, create symbols of their relationship (the next version of the product, the strategic plan, the funding plan) which help shape subsequent behaviour and circumscribe deviant behaviour. Entrepreneuring is about creating and orchestrating such rituals, taking on a symbolic role at the same time as being very active.

# 3 The Technology Development Process

A company was established to commercialize a specific cancer detection device, which was designed to carry out non-invasive, in-vivo detection of cancerous and pre-cancerous lesions, and more recently to treat cells it detected. The currently method used by clinicians is the analysis of a cell sample in a laboratory. The new technology reduces the time for diagnosis and improves the accuracy of detection from 50% to 84%. Other benefits are that nurses can perform the examination instead of doctors, and the improved accuracy should diminish the number of lawsuits served on clinicians for the lack of or the wrong detection of cancer.

The inventor of the technology was a university academic, who researched the area for over 9 years. He discovered that when a light was shone at a specific frequency, the reflection altered depending on whether the cells were abnormal or not. He created and empirically verified a bio-mathematical model which described how cells worked and built a machine that incorporated his two ideas. The business was incorporated in 2002, supported by his national bank's new venture investment company. Between 2002 and 2009 the inventor tried to get the market to accept his invention, without much success; only 6 machines were sold in all that time. Successive rounds of funding were acquired from the first investment company, and more recently an angel investor who believed in the technology. A CEO with more commercial experience was appointed by the board in 2007 (CEO1).

The product was being improved all the time; however the retail price was very high as each machine was custom made, thus restricting the market to clinicians who provided high end services. The operations struggled to fulfil even the small number of orders, because academic laboratories were being used to manufacture medical devices.

CEO1 established part of the company in the UK (the academic was based in another EU country), for three main reasons. The first was proximity to London, which helped win funding from a UK investment company; the second was the skilled labour force available which could help with the development and commercialisation; and the third was that medical devices are reputedly more readily acceptable by customers if they are designed and manufactured in Switzerland, Germany or the UK. CEO1 used informal networks to recruit a Finance Director (FD) and Operations Director (OD), both initially on consultancy bases, later becoming permanent. Both these people had extensive experience in large companies and more recently experience in new ventures. They found out about the job through informal contacts. In late 2009, after a considerable time battling with the inventor about trying to improve the technology, CEO1 decided to leave the company, and the OD took on role of CEO2 at the behest of the board. A Sales & Marketing Director (SMD) was also recruited using informal networks.

The original inventor is still on the board of the company, but has resisted successive attempts to change the technology configuration for commercial purposes. He has what might be termed satellite status in the company, being given resources to develop technology in a laboratory in his home country with a small number of staff, still passionate about his inventions.

CEO2 has over 20 years of experience of Research and Development with a multinational consumer electronics company, and latterly a medical device SME. He was originally tasked with understanding the poor product performance in the market place at the time. To do this, together with the newly appointed Sales and Marketing Director, they have actively sought to understand the medical practitioners' requirements and incorporate that into the technology.

During a fact finding mission to the USA in March 2009, CEO2 identified 3 segments in the prospective market, and decided there needed to be versions of the device developed for each tier, through establishing a platform and modular variations. Recognising the importance of FDA approval for the further commercial success of the product, it was filed for in August 2009 with assistance from US based regulative consultants. The idea was to position the technology as a variation of an existing technology, in order to smooth approval.

With the board's active participation (they visited and met the company chosen), a 3rd party R&D and manufacturing company was selected to help improve the technology, with a clear remit. The brief required the selected company not to change any of the essential components, the Intellectual Property; but they were tasked to focus on the following elements: the design had to be smaller, lighter and ergonomic; pass all regulatory requirements for the target markets and be manufactured at half the price.

Two potential subcontractors were shortlisted and invited to quote for the business. CEO2 visited each company. The subcontractor selected is a UK-based medical and laboratory device designer and manufacturer. CEO2 had experience of working with the chosen subcontractor before, and knew they could cover both design and manufacture, thought they had a passion for engineering but also he understood their limitations.

The company has around 80 staff at its UK site, and the ability to access manufacturing capacity in China. It has over 30 years' experience, and prides itself on being accredited to BSEN ISO 9001: 2008, BSEN ISO 13485:2005 and FDA QAS CFR 21 part 820. The subcontractor thought that its skill in engineering knowledge, project management, standards of working and ability to progress customer devices through the classification process was key to allowing them to remain in business when much UK manufacturing has disappeared. The company operates its own version of a stage-gate product development process to manage customer projects, and will accept new business entering or leaving the process at any stage, subject to the establishment of firm requirements on the scope and detail of the project, and everyone has to take stock if these are changed.

Each company had its own perspective on what needed to be done to develop the technology, with the main overlap being to make it more acceptable for customers. When we talked to stakeholders in the main company, the strategic objectives cited above were stressed; when we talked to the subcontractor stakeholders, a new emphasis, which had been mentioned before but not fore-grounded, emerged. The subcontractor was happy to work with the rough concepts provided by the customer company but was concerned with which standards categorisation the medical device would belong to, as they believed the one it was in was hampering customer perceptions. The internal workings of the technology had to be altered in order to meet the standards for the higher classification. The electrical subcomponents in the device had to be 'earthed' for safe operating, and the subcontractor thought the external notification body would not qualify it if modifications were not made.

Both companies were concerned with the ergonomics of the machine, and the subcontractor created a mock up medical setting with a dummy; and in conjunction with CEO2 & SMD held discussions with the customer and clinicians that were already using the previous version. They came up with some new concepts, developed non-functional models, then revisited the clinicians who were used to a high standard of kit and they all said that they liked it. Keeping the device mechanically stable was essential to the improved detection figures, and in the engineering prototype devising a braking device was a challenge, which CEO2 was particularly keen to get right. The subcontractor company engineer had an epiphany – whilst travelling in a cab he looked out the window at a bicycle and thought perhaps that mechanism would work. "Sometimes the solutions are thrown at you rather than hard won."

The development involved several meetings between the CEO2, SMD, and the team of mainly 3 engineers and quality managers in the subcontractor. The CFO and other staff from the subcontractor were incorporated in the discussions when necessary. The meetings were punctuated by visits to clinicians for verification that the development was going in the right direction and board meetings, which were important hurdles in the continuing funding and existence of the company and technology. For example a key meeting from CEO2's perspective took place at the subcontractor company, as he wanted to reinforce the changes of direction in the company and technology and the faith in the subcontractor by letting the board members see the subcontractor and meet the people first hand. Although one of the key people could not be there at the meeting, it went as CEO2 intended, and the board are bought in to the current direction. As of March 2011, FDA approval has been received, and the company has orders for 10 devices to be manufactured.

# 4 Discussion

The following section relates our case material to the ideas described in the earlier section on previous writing on the area. It can be argued that there are several strands of 'effectuation' (Sarasvathy, 2001) taking place, all of which involve CEO2, but none exclusively him. These collectively comprise a 'worknet' (Latour, 2010) with some nodes being included and some not for the various activities, and the important feature is the relationship between the nodes or actors, with some weaker relations effectuating better. Seven possible strands are 1) the invention and turning into a device of the technologies 2) the creation of the business venture as an organization 3) the funding stream creation and expansion 4) the building of relationships with the medical profession for testing and market knowledge 5) the creation of a distribution network, 6) the pursuit of certification and categorization and 7) and most pertinently for our paper, the creation and development of the relationship with the subcontractor design and manufacturing company.

We would like to say that all the actors in the network have at some point performed the arbitrage role identified by the early authors described in Van Praag (1999), therefore to some extent they are all 'entrepreneuring'. The main people in the network we have interviewed have all had knowledge acquired over extensive careers. There was engineering knowledge predominance in CEO2 and the subcontractor company, which was supported by the business knowledge of the CFO, SMD and board. Examining the prior experiences of the people highlights that none are career entrepreneurs as the cases were in the Howarth et al paper (2005). Nor can the company be said to be motivated by creating a private kingdom – there have been three leaders and the way the current CEO and CFO became part of the company shows how weaker filaments (Latour 2010) have become stronger through time and mutual understanding and the energy developed through the interactions.

There is a deep sense of security transmitted about the basic technologies and the new version of the device; which seems contradictory given that this is a start up venture seeking further funding to maintain its existence whilst the new version of the machine is marketed. The basic intellectual property has always been sound, and there is deep emotional attachment (Collins, 2004) to trying to commercialize this product because it could actually save lives. This gives the narratives a degree of certainty beyond if the product were a consumer good – something that the previous literature does not appear to address. Fletcher's paper (2006) is very resonant, as there has been a huge amount of effort, repeated negotiations with many groups and constant dialogues; however this has not been taking place all along the process. If we were to discuss the moments of translation (Callon, 1986), it is possible to see that the problematisation phase was successfully achieved by the inventor, but the translation process stalled at that point – the entrepreneuring was unsuccessful. CEO2 in a far shorter period of time has moved the process through more phases eg interessement and enrolment of a far wider network. CEO1 managed to question the dominance of the inventor, but not loosen it, which CEO2 managed through a set of meetings, which we would argue are business rituals, such as ones with the board establishing the market strategy and subcontractor relationship. None of this could have been achieved without the research on the clinical process; rituals (Collins, 2004) the inventor was not prepared to do or act on the findings. It is not possible yet for us to argue that full translation has occurred because of the stage the company is in the technology development process – however we would like to record the 'collective effervescence' (Collins 2004) of the stakeholders we interviewed and note how many of the moments of translation which have taken place so far have been embodied – there has been a huge importance in physical meetings which have been closed to outsiders, engagement in the task and all the people concerned are interested in developing the technology and there is a shared mood with one notable exception. The progressive side-lining of this person and termination of that version of the machine has given momentum and solidarity to the more extensive network which seems now more capable of entrepreneuring successfully.

# 5 Conclusions

In this paper we have reviewed theories of entrepreneuring and accepted Steyaert (2007) and Howarth et al's (2005) invitation to use other paradigm theories to analyse processes of entrepreneuring surrounding the development of a cancer detection and treatment technology. Combining ANT with Collin's work on Interaction Rituals we argue that entrepreneuring takes place in a worknet of many people performing simultaneously many strands of activities, in no way a linear discrete process overall, although containing short sequences of linearity. We would like to stress the openness of the network and the power of interactions which can mobilise positive and negative emotions, and these are all contributing to the processes of entrepreneuring.

# References

Alvesson, M. and Deetz, S. (2000), Doing Critical Management Research, Sage, London.

Burrell, G. and Morgan, G. (1979), Sociological Paradigms and Organisational Analysis, Heinemann Educational Books, London.

Callon, M. (1986) Some Elements of a Sociology of Translation, in Law, J. (1986) Power Action and Belief: A New Sociology of Knowledge, Sociological Review Monograph, no. 32, London: Routledge and Kegan Paul, pp 196 - 233

Collins, R (2004) Interaction Ritual Chains, Princeton University Press, Princeton.

Fletcher, D.E., (2006) Entrepreneurial processes and the social construction of opportunity, Entrepreneurship & Regional Development, 18 September (2006) pp 421-440

Deleuze, G. and Guattari, F. (1987) A Thousand Plateaux, University of Minnesota Press, Minneapolis.

Gartner W.B. (1988). Who is an entrepreneur? Is the wrong question.Entrepreneurship: Theory and Practice, Summer 1989 Vol 13, Issue 4 pp 47- 68

Goss, D. (2005) Schumpeter's Legacy? Interaction and Emotions in the Sociology of Entrepreneurship, Entrepreneurship: Theory and Practice, March 2005, Vol 29, Issue 2 pp 205-218

Howarth, C., Tempest, S., and Coupland, C., (2005) Rethinking entrepreneurship methodology and definitions of the entrepreneur, Journal of Small Business and Enterprise Development, Vol. 12 No. 1, pp. 24-40

Latour, B. (1986) The Powers of Association in Law, J. (1986) Power Action and Belief, a New Sociology of Knowledge?, Routledge and Kegan Paul, London, pp 264 - 280

Latour, B (2010) "Networks, Societies, Spheres: Reflections of an Actor-- Network Theorist" Keynote speech for the International Seminar on Network Theory: network Multidimentionality in the Digital Age, Annenberg School for Communication and Journalism, Los Angeles

Low M.B. and Abrahamson E. (1997). Movements, bandwagons, and clones: industry evolution and the entrepreneurial process. Journal of Business Venturing, Nov 97, Vol12 Issue 6 p 435- 457

MacMillan, I. (1986) To Really Learn About Entrepreneurship, Let's Study Habitual Entrepreneurs, Journal of Business Venturing, 1, pp 241-243

Marshall, A. (1930), Principles of Economics, Macmillan, London - First edition 1890.

McMullen J.S., and Shepherd D.A., (2006) Entrepreneurial Action and the Role of Uncertainty in the theory of the Entrepreneur, Academy of Management Review, Vol. 31, No. 1, 132–152.

Reynolds P. D. (1991). Sociology and entrepreneurship: concepts and contributions. Entrepreneurship: Theory and Practice. Winter 91, Vol 16, Issue 2, pp 47- 70

Sarasvathy, S. D. (2001). Causation and effectuation: Toward a theoretical shift from economic inevitability to entrepreneurial contingency. Academy of Management Review, Vol 26, Issue 2, pp 243-263

Schumpeter, J. (1934) The Theory of Economic Development, Harvard University Press, Cambridge Mass.

Steyaert C, and Katz, J (2004) Reclaiming the Space of Entrepreneurship in Society: Geographical, Discursive and Social Dimensions, Entrepreneurship and Regional Development, 16, May, pp 179–196

Steyaert C, (2007) 'Entrepreneuring' as a conceptual attractor? A review of process theories in 20 years of entrepreneurship studies, Entrepreneurship & Regional Development, 19 November 2007, 453-477

Thornton, P (1999) The Sociology of Entrepreneurship, Annual Review of Sociology, 1999, Vol. 25 Issue 1, p 19 – 47

Van Praag, C. M (1999) Some Classic Views on Entrepreneurship. De Economist Sep 99, Vol.147 Issue 3, p 311

# New Product Development and Commercialisation Process in the SME Fashion Design Houses

**Shuyu Lin and Dr Niall Piercy**
School of Management, University of Bath, Bath, UK

**Editorial commentary**

The focus of Lin and Piercy's case is on the process of new product development in SMEs in the UK fashion industry, a highly competitive and rapidly changing industry. Specifically, they explore the knowledge and skills that a fashion designer must possess in order to contribute to the success of new product development, noting in particular that both creative and entrepreneurial competencies are required if commercialisation of products is to be achieved smoothly.

The authors construct a conceptual framework to illustrate the antecedents to the new product development (NPD) process, the competencies fashion designers require, and the evaluation of performance of the NPD process.

Some points for discussion and learning arising from this case include:

- The range of knowledge and skills required to achieve success in the new product development process.
- The education and training needs of fashion designers.
- The balance between creativity and competitiveness.
- The entrepreneurial process and the entrepreneurial personality.

- The place and contribution of mixed skill teams in the new product development process.

---

**Abstract**: This paper concentrates on fashion designers' competencies in small and medium-sized enterprise (SME) fashion design houses. Specifically how they respond to the highly uncertain, extremely competitive, and the fast-changing fashion context when undertaking new product development. Focusing on integrative and dynamic competencies, it considers designers' impact on a firm's performance in the new product development (NPD) process. Designers must play a dual role as creative individuals and as entrepreneurs providing a complicated situation. This paper attempts to frame a problem in the UK designer fashion industry where British based freelance designers have been praised for high levels of creativity but criticised for their lack of business sense.

In the decade, UK government and the market environment have reputations to promote creativity on the following evidence. The Department of Trade and Industry has been more active since 1997 in promoting design and innovation. Also, New Generation design bursaries, funded by Topshop since 1999 and chosen by a rotating panel of key industry figures in collaboration with the British Fashion Council, enable designers to fund the critical presentation of their collections to press and buyers at London Fashion Week. Also, London is a world-renowned centre for young talents and design education with world-class academic institutions such as Central Saint Martin's College of Art and Design.

Having explored and analysed the existing problem and potential opportunities in the UK designer fashion industry, this paper determines where independent designers lack specific competencies and the competencies required for successful new product development, namely, technological competencies, commercial competencies, market sensing capabilities, strategic mindsets, project management skills, and networking skills. These competencies are potential determinants which influence both financial performance such as sales and non-financial performance such as innovation. Finally, after gaining clearer insights into the NPD process where a designer plays multi-dimension roles as an artist and as a manager, managerial implications for the designer fashion industry are presented.

**Keywords**: competencies, designer fashion, dynamic capabilities, innovation, new product development, small and medium-sized enterprises

# 1 Introduction

Unlike large businesses, fashion design houses are relatively lean and small in structure particularly in the UK. Designers convert their design ideas and concepts to products in the NPD process with one or two assistant in their own workshops or studio. In this case, a designer's competencies are at the centre of contribution to the performance. A designer is the one who transforms his/her intangible talents and individual intellectual to solid seasonal fashion collections. A designer infuses the designer label with life.

Given that, this paper explores the key competencies of a fashion designer in the new product development. The emphasis is on the following aspects: technological competencies, commercial competencies, market sensing capabilities, strategic mindsets, project management skills, and networking skills. An understanding of these key competencies may not only improve the process of design innovation, but may also increase the returns from these innovations through enhanced integration of the competencies.

The aim of the paper is to develop a theoretical framework. By doing so, the UK designer fashion context is analysed first. The NPD process from a competence integration perspective is presented. Then the NPD process particularly in the SME fashion design houses is discussed. Finally, the framework is introduced along with managerial implications.

# 2 The context of the UK designer fashion industry

The fashion industry is a mature market characterised by increased competition and price deflation due to overcapacity (Parrish et al., 2006). The level of the fashion industry spans from haute couture fashion houses to mass manufacturing high street retailers. In this supply chain, high-end designer labels are upstream fashion trend setters; by contrast, high street retailers are downstream trend adapters who mass manufacture similar trendy collections originated from runway fashion shows (Garner and Keiser, 2003).

Designerwear has been received considerable attention recently partly due to the rise of luxury interests and consumption. Spending on both women's and men's designerwear has grown steadily in the past few years (Mintel, 2006). Designerwear refers to designer-led collections such as haute couture and mainline designer ready-to-wear. In comparison with haute couture, designer ready-to-wear is at the second fashion level in terms of price and more profit-oriented (Keiser and Garner, 2003). Although relatively less expensive than haute couture, it is impeccably made of the fine and luxury fabrics with limited quantity.

Fashion designers are varied accordingly. In this paper, fashion designers are those who hold their bi-annual fashion shows or exhibitions in the world-leading fashion capitals such as London, Paris, Milan, and New York. Neither mainstream designer ranges for high-street retailers like designers at UK department store Debenhams nor crossover corporation like designer limited collections in global high street retailers H&M are addressed in this paper. It focuses on the UK-based designer ready-to-wear segment particularly in the UK fashion industry.

The UK designer fashion industry is the fourth largest in the world after the US, Italy, and France (Department for Culture, Media and Sport 2001). London, New York, Paris and Milan, fashion capitals in those countries, hold major fashion shows where the latest designs are produced and exhibited. London, as a major player on the world's fashion stage and leading force in the UK industry, must compete with those three other cities. However, compared with the US, French, and Italian industries that are characterised by large multinational brands, the UK industry is different in structure because it mainly consists of small companies (Department for Culture, Media and Sport 2001). A large number of designer labels come under the umbrella of the major conglomerate like world leading prestigious fashion company LVMH and Gucci in France and Italy respectively.

Creativity friendly environment, rather than commercial dominant market, is another characteristic of UK designer fashion industry. However, creativity solely is difficult to protect businesses to be invulnerable in the competitive marketplace. In recent years, however, London's leading status for fashion has been challenged. The UK domestic designer market has the proven creative skills for designer fashion, but not the business acumen to reap the profits (Management Today 2004).

The conditions in any fashion industry make these circumstances especially paramount; that is, because the industry is fundamentally 'creative'. This means it has its origins and is driven by individual creativity, skill and talent, along with the potential for wealth and job creation through the generation and exploitation of intellectual property, individual creativity, and intellectual capital (Department for Culture, Media and Sport 1998; Jones et al, 2004). In addition, due to the characteristics of the fashion industry of having short product life cycles, high volatility, low predictability, and high impulse buying, the fashion market is highly competitive and especially challenging to manage (Christopher et al., 2004). It is, therefore, vital to find the right balance of efficient and effective management as well as space for innovation and creativity.

## 3 The evolution of NPD process

The research of NPD process can be traced back to the Phased Project Planning developed by NASA in 1960s (Von Stamm, 2003). The process originally applied to complex and large scale projects. Four sequential phases were identified, namely: preliminary analysis, definition, design, and operations. In this vein, following research explored NPD process on the linear process construct. Along with extensive efforts, key phases in the NPD process were refined which are identification of the need, concept generation, design and development, production, and launch (Cooper, 1988; Perks et al., 2005).

Linear product development model has been criticised to be over generalised and unable to fit the increasingly complex business environment (Takeuchi and Nonaka, 1986). The emphasis of NPD research, therefore, has been shifted from effectiveness to efficiency to respond to the high speed and fast changing business environment (Cooper, 1994). Recent research suggests that new product development should be a more integrated and dynamic process having impacts on long-term success of a company (Cooper, 1994). In this vein, team-based cross-functional approach has been developed and has attempted to address the interaction issues within different departments in an organisation (Hart, 1993). Focusing on speed, flexibility, and organisational learning is key evolution of NPD process (Cooper, 1994; Takeuchi and Nonaka, 1986).

Following this research stream, numerous new product development approaches have been developed based on cross-functional perspectives. Successful new product development is based on the integration of multiple functions mainly operations and marketing flowing through the process (Cooper, 1994; Marsh and Stock, 2003).

# 4 New product development process from a knowledge-based perspective

Team-based cross-functional model drawn by the large business context, however, is unlikely to fit SMEs which is of lean human resources. Alternatively, knowledge-based perspectives shed a fresh light on the NPD research (Trott, 2005).

Unlike large firms, SMEs have been viewed highly for their innovation but have had lean resources, particularly technological competencies, to protect innovation activities in the NPD process. Innovation capabilities working through new product development in SMEs are significantly related to growth and profit performance (Wolff and Pett, 2006).

SMEs, like large firms, attempt to develop and combine technological and commercial competencies (Tidd et al., 2005). Compared to large firms' new product development, SMEs have strength in communication, greater contribution to innovation in certain sectors like creative industries, and speed of decision-making (Tidd et al., 2005). However, technological disadvantages hold firms back to develop and manage complex and long-term projects.

Specifically, the scenario mentioned in the previous section indicates that designers' intellectual property is the core asset of the business. It is a driving force throughout the entire business. Knowledge defined at an individual level has been accumulated through the projects from ideas, production, to launch. This knowledge turns into the intellectual asset of the following product development, and in turn, into long-term competencies of an organisation. According to the definition of dynamic capabilities, difficult-to-replicate competencies at the individual level may provide a fresh thinking of development of competitive advantage in the fashion design houses. Effective integration of capabilities becomes an alternative thought (Marsh and Stock, 2003). Consequently, this paper defined competitive advantage as a designer's individual competencies which create collection portfolio in this specific context.

Knowledge-based dynamic integration process in product development is defined as the process of collecting, interpreting, and internalising technological and marketing capabilities from past new product development projects and incorporating the knowledge in a systematic and purposeful manner into the development of future new product (Marsh and Stock, 2003). Technological capabilities are physical components from idea generation, product design to manufacturing. Activities within the process includes commercialisation capabilities, as known for marketing competencies, are built centrally upon the understanding of customers' needs.

# 5 NPD in SME fashion design houses

As earlier mentioned, the fashion industry is fundamentally creative and competitive. Fashion design houses must identify their competitive advantages to survive in this highly competitive marketplace. At the heart of a firm's competitiveness are its core capabilities or competencies (Teece and Pisano, 1994). These competencies are firm-specific, built up over time from day-to-day activities and experience, and difficult to be replicated by competitors.

Fashion trends are the premise of the NPD in the designer fashion industry. The definition of the fashion trends can be analysed in two folds. First, they are design ideas which major collections have in common (Mete, 2006). That is, each designer collection is basic units of the fashion trends which are usually categorised and reported by fashion press each season. Meanwhile, fashion trends are the appearance and construction of fashion products which relate to a particular season (Jackson, 2007).

Early research on fashion trends suggested that they affirmed themselves in a spontaneous and uncoordinated way (e.g., Blumer, 1969; Schrank and Gilmore, 1973). However, recent research has shown that fashion trends are the result of intentional and coordinated efforts by actors such as designers and senior management executives through collective and focused actions deliberately shaping the fashion industry and its paths of innovation (Rinallo and Golfetto, 2006). Fashion trends are in fact a few years in the making, carefully nurtured, and managed to meet customer tastes by the time the products reach the market.

The NPD process in fashion involves several key phases from initial design concept to final production. The activities in the process involve sourcing inspirations, analysing trends in fabrics, colours and shapes, setting the theme of the collection, developing sample garments, and finally selecting the best attributes of collections to then implement (Learned, 1969; Ansoff, 1965).

The first step in creating a design is to research and understand the current market and then to forecast potential fashion trends. In this case, fashion research and forecasting as well as creativity are at the centre of initial activities of fashion process (Davis, 1992). Fashion research may starts from analysing trends in fabrics, colours and shapes. Inspiration is likely to emerge in designers head during the research. Sources of inspiration and personal interpretation are key determinants of creativity (Mete, 2006). There are various ways how designers are inspired. That is, designers are likely to be inspired at any time under any circumstances. It could be the history, contemporary art, social scenario, natural objects, and even their earlier collections (Mete, 2006).

In the early stage in the fashion NPD process, designers' main tasks are focusing on visual design. Fashion illustration is an approach to visualise ideas from inspiring sources. Technological skills like sketching and personal interpretation have direct impacts on innovation and creativity. Sketching is a start point for designers to translate designers' intangible thoughts and particularly emotions into physical garments. Then, designers examine their sketching books and refine their ideas into the theme of the collection.

Once the theme has been chosen, a prototype is created and then tried on a model to see whether adjustments are needed. Material finding is the following step to make those illustrations real. In this stage, all designers need to do is to search for fabrics which match to the theme set previously. Technological design skills such as trimming and sewing are important in the garment construction. In this case, prototyping assists designers to narrow down their design ranges and to select final designs to be made for collection for sale. Finally, designers launch their entire new collections in fashion shows.

Depending on the size of a design house, fashion designers may have varying levels of involvement in design and production aspects (Perks et al., 2005). Different from those working in big design houses with various departments, SME fashion designers are required dynamic capabilities to carry out multiple tasks in the NPD process. They usually perform most of the technological roles like designing the clothing, patternmaking, and sewing tasks; in addition, holding launch events on their own rights.

# 6 The framework of designer fashion NPD process

After the analysis of the NPD process particularly in the fashion industry, the conceptual framework is presented below. The structure of the conceptual framework is based on the NPD process which converts resources into products. As designers undertake leader roles, their capabilities become the main contribution to inputs both for the product and the process. The conceptual framework, introduced in Figure 1, is threefold. The first component provides antecedents of the innovative NPD process. A second component introduces elemental competencies designers required in dealing with new product development. The third component evaluates performance which generate from the NPD process. The set of elemental competencies at the centre of the NPD process framework is what this paper focuses on. The framework aims at integrating dynamic capabilities so that designers can adapt to the fast changing environment.

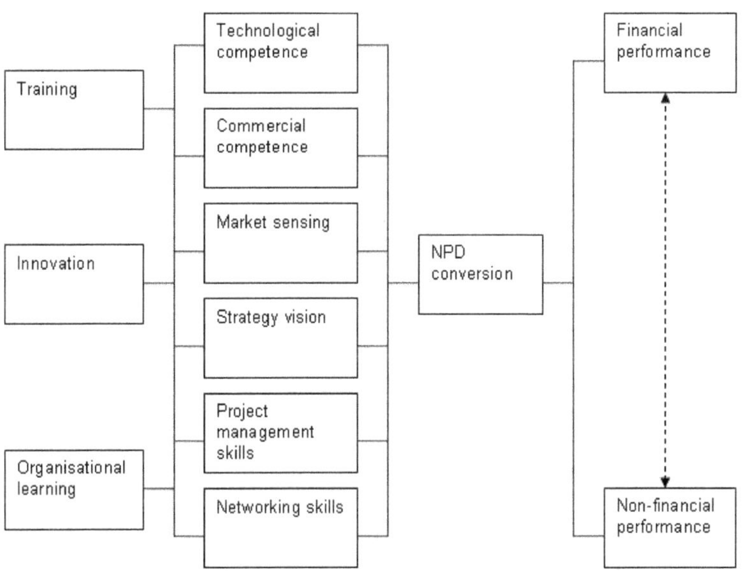


**Figure 1:** The framework of the NPD and commercialisation process

The literature in the previous section makes it possible to identify dynamic capabilities which are conceptualised below: technological competencies, commercial competencies, market sensing capabilities, strategic mindsets, project management skills, and networking skills. The framework illustrates the key competencies seen as inputs cause effects on the organisational performance, seen as outputs via the NPD conversion within the process. The main contribution of the framework is the integration of competencies and also the space both for innovation and commercialisation. In this way, it can be ensured that organisational learning and knowledge transformation continuously adjusts and improves the capability for change adaptation in the fast pace external fashion environment.

## 6.1 Antecedents of the NPD process

In order to ensure the capability of competence integration, three of the antecedents should be addresses in advance. Without these, each NPD process is likely to become a one-off project. As a result, SME fashion design houses are unable to accumulate their efforts and develop their non-replicated capabilities which can turn out to be future competitive advantage to sustain in the marketplace.

### *6.1.1 Training*

There are generic barriers to integration and cooperation of the characteristics of marketing and design functions (Grønhaug and Kristensen, 2007). At the individual level, a designer educated in an art school tends to work with visual and material expressions. By contrast, marketing knowledge has more decision attitude which expresses in verbal, qualitative and quantitative models (Bolan and Collopy, 2004). Successful designers must have dynamic capabilities to integrate marketing and design. There are various ways to integrate marketing and design. According to the characteristics of designer fashion houses, individual learning and training is likely to provide designers such knowledge and skills. As a result, design turns out to be a strategic asset of a designer label (Grønhaug and Kristensen, 2007).

### *6.1.2 Innovation*

A growingly competitive environment drives firms to place more emphasis on creativity and innovation as the core value of the business (Perks et al., 2005; Von Stamm, 2003). Innovation can be viewed as creativity plus commercialisation (Von Stamm, 2003). In addition, it is about knowledge which means creating new competencies and refining existing capabilities through combining different knowledge set (Tidd et al., 2003). The knowledge is generated from the experience of the previous projects and new inspiration from external stimuli.

Creativity is the core competence in design. Commercialisation is the way of enticing customers' desires. In this case, innovation in the designer fashion industry can be in the form that the designer collection is of creativity and allure at the same time. As earlier mentioned, fashion trends have been well-designed by designers before entering the market. There is a strong correlation between the performance and the competencies of designers to allure customers by their collections. For example, to be a truly successful designer like Coco Chanel, designers must have the instinct and ability to sense and create what women want a split second before they know it themselves (Davis, 1992).

### *6.1.3 Organisational learning*

Organisational learning is important to re-innovation of a firm (Rothwell and Gardiner, 1988). Learning means an organisation captures lessons from both success and failure which benefits the following projects (Tidd et al., 2003). The willingness of learning is essential in this stage.

Learning is defined as a continuous changing in capability which is applied to an organisation (Kelliher and Henderson, 2006; Leonard-Barton, 1992; Rothwell and Gardiner, 1988). Learning can be at the technological level which means adding knowledge to technological competencies. Also, it can occur at the managerial level which strengthens the capabilities needed for effective product innovation management (Tidd et al., 2003). However, organisational learning in SMEs is different from large companies.

Small businesses are struggling with organisational learning due to resource poverty such as financial constrains and informal management process (Kelliher and Henderson, 2006). In SME fashion design houses, designers are at the centre of the decision making process and they use their intuition rather than through a formal decision procedure. In this case, individual learning is a key knowledge factor (Kelliher and Henderson, 2006).

## 6.2 Competencies of fashion designers in the NPD process

Core competencies are defined as a set of differentiated skills and complementary assets which provide the basis of a firm's competitive advantage (Teece et al., 1997). From a knowledge-based perspective, knowledge and skills embodied in designers is the one most associated with core competencies (Leonard-Barton, 1992). In addition, individual knowledge and skills are at the centre of new product development. In this section, core competencies required for successful NPD are illustrated below.

### *6.2.1 Technological competencies*

The NPD process consists of creative individuals, firms operating functions, and external linkage of firms such as market needs and market changes lead to emerging opportunities (Von Stamm, 2003). Technological competencies, defined as physical components from idea generation, product design to manufacturing, are likely to have impacts on non-financial performance such as innovation, organisational learning, and tacit knowledge transformation. It is the intangible asset for fashion design houses to the following product development process. By contrast, commercial competencies are likely to influence the financial performance particularly on sales and profits.

In the case of more mature and established products like fashion, competitive profit growth is not generated simply from being able to offer competitive prices but also from a variety of non-price factors like design, customisation, and quality (Tidd et al., 2003). Design function is adopting a more prominent position in the management of the product development efforts. As a result, converting uniqueness of tacit talent into explicit and solid collections is at the top of a competence set designers have to develop.

### *6.2.2 Commercial competencies*

Apart from design activities, designers have full autonomy and undertake multiple roles in the NPD process. Business and marketing knowledge for strategic decision making are also required. Non-design knowledge like fashion awareness as well as fashion research and forecasting play important roles in the early stage of the NPD process. Specifically, fashion designers are expected to understand the market such as consumer preferences and sales statistics in the previous seasons (Jackson, 2007). Transferring customer knowledge into new product design is another issue to produce market-oriented fashion collections. As designer fashion industry is described as a market-driving business environment, commercial competencies to drive the market are crucial to business survival. Most important, they are how designer labels win reputations as market leaders in terms of innovation.

### 6.2.3 Market sensing

In fashion, design is related to the quality and beauty rather than functionality. Thus, design is more concerned with the understanding of something which has not existed and with the delivery of an imaginary state (Grønhaug and Kristensen, 2007). In this case, market sensing capability of a designer is at the central of value adding (Schindehutte et al., 2008). Differently, designers attempt to create a new market and drive a new taste rather than respond to the exiting market. In this situation, designers have sensed customers' needs and desires before customers recognise and articulate themselves. That is, designers are educators who teach customers and, by contrast, customers are learners who are taught new preferences. Given that, market sensing capabilities are deemed as opportunity discovery capabilities from a strategic perspective. These capabilities enable a designer fashion house to anticipate new needs, to envision marketing offerings and to acquire, develop and create the required new resources associated with this innovation, which result in a sustainable competitive advantage and enduring superior financial performance (Schindehutte et al., 2008).

Market sensing capabilities are rested on innovation which drives designers to come up with new design concepts. Innovation provides each fashion product with a distinctive mix of attributes. Invention and creativity are key elements at the first stage in the fashion trend development process from creation to consumption (Sprole, 1985). A fashion product with the right attributes can become an icon which satisfies customer needs and ignites their desires. A mixture of attributes like colour, fabric, print, silhouette, styling detail, and trim in each fashion product reflect a very influential fashion trend (Jackson, 2007). Many pretentious designers fail because they lack innovation and sense of consumer needs; even successful designers risk failure in their attempts to capture a proportion of the consumer taste (Hollander, 1993).

### 6.2.4 Strategy vision

New product development is a strategic tool of a firm to capture and retain market shares, increase profitability, and differentiate with competitors (Tidd et al., 2003). Fashion designers that work in houses producing exclusive designerwear are viewed as creative directors rather than simply designers since they are also responsible for the strategic decision making of their own businesses (Jackson and Shaw, 2006; Perks et al., 2005). In other words, designers carry out leadership actions to manage the whole process. In the case of fashion design SMEs, designers are responsible for almost everything in their own design studio from generating design concept to producing the completed garment. Strategic mindsets are applied to the situation while fashion designers make decisions. For example, when setting the theme of the collection, they should consider how to differentiate with others and which concepts should be chosen.

### 6.2.5 Project management

In responding to environment and market changes, new product development projects become the focus of innovation and core competencies renew (Leonard-Barton, 1992). In this case of survival in the highly competitive and uncertain marketplace, managing each project by integrating competencies efficiently and effectively, as well as the space for continuous innovation in the future provides a competitive advantage. A set of project management skills of designers as process leaders are competencies which combine design, resource-based decision making, and commercialisation strategy decision. Designers are expected to undertake multiple activities to have better interactive and communication skills achieve better technological and commercial integration. The better interactive within dynamic capabilities, the better performance would be achieved at the end of the process. In addition, project integration capabilities are expected to take place in each project and enable to make sure the space for either individual or organisational learning (Grønhaug and Kristensen, 2007).

### 6.2.6 Networking skills

Fashion week events are trade fairs for designers to network with potential buyers and the press. In this case, designers hosting bi-annual events either catwalks or exhibitions in the fashion week are expected to have networking capability to present their collection, communicate with those important people, pursue and convince potential buyers to stock in the collections during the show. Without those social skills, even the greatest collections are unlikely to standout in the overwhelming fashion fair. Also, designers are expected to communicate with upstream textile manufacturers for material sourcing.

## 6.3 Performance

The existing research suggests a strong correlation between market performance and new product development (Tidd et al., 2003). However, to what extent does a firm's core competencies contribute to its success in the NPD process is difficult to measure. Alternatively, success is assessed from a performance perspective (Cooper and Kleinschmidt, 1991; Wolff and Pett, 2006). Performance is a complex and multi-dimensional construct which may not be assessed by financial performance only (Hart, 1993). Apart from financial performance measured by profits, revenues, and turnover of a company, non-financial success is related to design, innovativeness, design, technological and commercial capabilities, and proficiency of activities. In this vein, the paper discusses on both the non-financial achievement as well as revenues a business generates.

A firm's success is built up on three levels, namely, product level, project level, and company level. A product-level achievement is the first step, followed by project level success. It's clear to say that a firm-level achievement is based on the series of successes at product and project level (Cooper and Kleinschmidt, 1995). In other words, strategically, integrating dynamic capabilities to sustain in the changing and highly competitive environment through the NPD process over time is far more crucial than a success of single products in a specific period of time.

# 7 Conclusion and managerial implications

In this paper, a theoretical framework drawn from the literature review for identifying key competencies of the integration for new product development and commercialisation process has been proposed. By doing so, the technological, commercial, and individual intellectual perspectives are adopted to identify key competencies particularly for fashion designers in SME fashion design houses. As shown in the previous sections, the framework introduced in Figure 1 provides a picture of possibilities successful new collection could be made. However, to be practical, more empirical research is required to validate the framework and test its applicability in designer fashion houses.

## References:


Ansoff, H. I. (1965) Corporate strategy; an analytic approach to business policy for growth and expansion, New York: McGraw-Hill.

Blumer, H. (1969) "Fashion: From Class Differentiation to Collective Selection", The Sociological Quarterly, Vol.10, No. 3, pp 275–291.

Bolan, R.J. and Collopy, F. (2004) "Design matters for Management". In: Bolan, R.J. and Collopy, F., Management as Designing, Stanford, California: Stanford Business Books.

Borrelli, L. (2008) Fashion Illustration by Fashion Designers, London:Thames & Hudson.

Christopher, M., Lowson, R. & Peck, H. (2004) "Creating agile supply chains in the fashion industry", International Journal of Retail & Distribution Management, Vol. 32, No. 6, pp 367-376.

Cooper, R.G (1994) "Perspective third-generation new product development", Journal of Product Innovation Management, Vol. 11, No. 1, pp 3-14.

Cooper, R.G. (1988) "The new product process: a decision guide for management", Journal of Marketing Management, Vol. 3, No. 3, pp 238-255.

Cooper, R.G. and Kleinschmidt, E.J. (1991) "The impact of product innovativeness on performance", Journal of Product Innovation Management, Vol. 8, No. 4, pp 240-251.

Cooper, R.G. and Kleinschmidt, E.J. (1995) "Benchmarking the firm's critical success factors in new product development", Journal of Product Innovation Management, Vol. 12, No. 5, pp 374-391.

Davis, F. (1992) Fashion, culture, and identity, Chicago: University of Chicago Press.

Department for culture, media and sport (1998) Creative Industries Mapping Document, London: DCMS.

Department for culture, media and sport (2001) Creative Industries Mapping Document, London: DCMS.

Hart, S. (1993) "Dimensions of success in new product development: exploratory investigation", Journal of Marketing Manegement, Vol. 9, No.1, pp23-41.

Hollander, A. (1993) Seeing through clothes, Berkeley: University of California Press.

Jackson, T. & Shaw, D. (2006) The fashion handbook, London: Routledge.

Jackson, T. (2007) The process of fashion trend development leasing to a season, in Hines, T. & Bruce, M. Fashion marketing: contemporary issues, Oxford: Butterworth-Heinemann.

Jones, P., Comfort, D., Eastwood, I. & Hillier, D. (2004) "Creative industries: economic contributions, management challenges and support initiatives", Management Research News, Vol. 27, No. 11, pp 134-145.

Keiser, S.J. and Garner, M.B. (2003). Beyond design: The synergy of apparel product development, New York: Fairchild.

Kelliher, F. and Henderson, J.B. (2006) "A learning framework for the small business environment", Journal of European Industrial Training, Vol. 30, No. 7, pp 512-518.

Kristensen, T. Grønhaug, K. and (2007) "Can design improve the performance of marketing management", Journal of Marketing Management, Vol. 23, No. 9-10, pp 815-827.

Learned, E.P. (1969) Business policy: text and cases, R. D. Irwin, Homewood.

Management today (2004) The fashion business: Can Britain cut it?, Management Today, 1 August, pp 39.

Marsh, S.J. and Stock, G.N. (2003) "Building dynamic capabilities in new product development through intertemporal integration", Journal of Product Innovation Management, Vol. 20, No. 2, pp 136-148.

Mete, F. (2006) "The creative role of sources of inspiration in clothing design", International Journal of Clothing Science and Technology, Vol. 18, No. 4, pp278-293.

Mintel (2006) "Designer Clothing Report", London: Mintel.

Parrish, E.D., Cassill, N.L. and Oxenham, W. (2006) "Niche market strategy for a mature marketplace", Marketing Intelligence & Planning, Vol. 24, No. 7, pp 694-707.

Perks, H., Cooper, R. and Jones, C. (2005) "Characterizing the role of design in new product development: An empirically derived taxonomy", Journal of Product Innovation Management, Vol. 22, No. 2, pp 111-127.

Porter, M.E. (1998) Competitive Strategy: Techniques for Analyzing Industries and Competitors, London: Free.

Rinallo, D. and Golfetto, F. (2006) "Representing markets: The shaping of fashion trends by French and Italian fabric companies", Industrial Marketing Management, Vol. 35, No. 7, pp 856-869.

Rothwell, R. and Gardiner, P. (1988) "Re-innovation and robust design: Producer and user benefits", Journal of Marketing Management, Vol. 3, No. 3, pp 372-387.

Schindehutte, M., Morris, M.H. and Kocak, A. (2008) "Understanding market-driving behaviour: The role of entrepreneurship", Journal of Small Business Management, Vol. 46, No. 1, pp 4-26.

Schrank, H.L. and Gilmore, D.L. (1973) "Correlates of Fashion Leadership:

Implications for Fashion Process Theory", The Sociological Quarterly, Vol. 14, No. 4, pp. 534-43.

Sproles, G.B. (1985) Behavioural Science Theories of Fashion, in Solomon, M.R. The Psychology of fashion, Lexington: Lexington Books.

Takeuchi, H. and Nonaka, I. (1986) "The new product development game", Harvard Business Review, Vol. 64, No. 1, pp 137–146.

Teece, D. and Pisano, G. (1994) "The dynamic capabilities of firms: an introduction", Industrial and Corporate Change, Vol, 3, No. 3, pp 537-556.

Teece, D.J., Pisano, G and Shuen, A. (1997) "Dynamic capabilities and strategic management", Strategic Management Journal, Vol. 18, No. 7, pp 509-533.

Tidd, J., Bessant, J. and Pavitt, K. (2005) Managing Innovation: Integrating Technological, Market and Organizational change, 3rd ed, Chichester: Wiley.

Trott, P. (2005) Innovation management and new product development, 3rd ed., Harlow: Financial Times Prentice Hall.

Von Stamm, B. (2003) Managing Innovation, Design and Creativity, Chichester: Wiley.

Wolff, J.A. and Pett, T.L. (2006) "Small-firm performance: modelling the role of product and process improvements", Journal of Small Business Management, Vol. 44, No. 2, pp 268-284.

# Regional Innovation and Competitiveness: Analysis of the Thessaloniki Metropolitan Region

**Panayiotis Ketikidis**[13], **Sotiris Zigiaris**[2] **and Nikos Zaharis**[3]
[1]CITY College – International Faculty of the University of Sheffield, Thessaloniki, Greece
[2]Aristotle University of Thessaloniki, Greece
[3]South East European Research Centre (SEERC), Thessaloniki, Greece

**Editorial commentary**

Ketikidis, Zigiaris and Zaharis discuss innovation in the context of regional economic development in the Thessaloniki area of Greece. Through an investigation of knowledge assets and knowledge flows, as well as the relationships between SMEs and SME support agencies, they identify the key role played by knowledge transfer as a catalyst for innovation at a regional level. The study concludes with a set of clear policy recommendations to help improve the innovation system in operation in the Thessaloniki region. These recommendations have wider application beyond this single metropolitan region of Greece.

This study was undertaken using semi-structured interviews with key players from industry, government support agencies, research laboratories and universities.

Some points for discussion and learning emerging from this study include:

- The keys to success in the knowledge transfer process.
- The roles of creative leadership, entrepreneurial acumen and entrepreneurial spirit in the economic development of a region.
- Adopting a multi-agency approach to research and development to improve regional innovation systems.
- The relationships between industry, government support agencies, research and technology development organisations, and universities in the creation of effective knowledge flows.

---

**Abstract**: The importance of innovation and knowledge-based policies in restructuring economies and increasing their competitiveness has repeatedly drawn the attention of policy makers and stakeholders. Regional policy is no exception to this trend. This paper attempts to map the demand and supply of Research and Development (R&D) in the metropolitan region of Thessaloniki and to draw a knowledge model that demonstrates the flow of knowledge in the region and the process of knowledge diffusion using the triple helix approach to depict the current situation and future dynamic emphasising issues that policy makers face. It analyzes Central Macedonia (CM) and East Macedonia and Thrace (EMTH) regional systems of knowledge flow and innovation, illustrating the institutional architecture and infrastructure associated with these systems. The data for this research was collected through in depth literature review and a number of in-depth interviews. Semi-structured face-to-face interviews were held with several key personnel within the research institutes, intermediaries and policy makers and triangulated with additional available information. The findings of this paper point to the conclusion that knowledge transfer is a priority and a starting point for regional initiatives regarding the increase in investment in R&D and innovative actions. The interest expressed by local agents to invest in knowledge transfer activities positively indicated the changing regional innovation environment. This optimism originates in a set of policy recommendations for the increase in knowledge investments, which were previously discussed widely in public and had been accepted as regional policy priorities by all local actors. The specific paper contributes to identifying and assessing knowledge assets and flows, as well as interactions of SMEs with support organizations, in the regions of CM and EMTH of Greece, and accordingly provides policy recommendations for the future economic development in the region.

**Keywords**: Innovation and regional development, innovation policy, research and development policies, knowledge economy, northern Greece

# 1 Introduction

The global importance of the concept of competitiveness has increased rapidly in recent years. Regions are increasingly seen as essential parts of the global society (Lundvall, 2007). Innovation evolved as part of sustainable development (Cooke, 2007) and became a driver of the competitiveness within the regions (Arundel et al, 2007). It was the early research of Porter (1990) that first defined national competitiveness as an outcome of a nation's ability to innovate in order to achieve, or maintain, an advantageous position over other nations in a number of key industrial sectors. According to Storper "area competitiveness" – at both national and regional level - is the capability of an economy to attract and maintain firms with stable or rising market shares in an activity, while maintaining stable or increasing standards of living for those who participate in it (Storper, 1997). Competitiveness involves the upgrading and economic development of all places together, rather than the improvement of one place at the expense of another. In addition, the knowledge-base of economies is an important measure of the creative capacity and investment in innovation, as well as its propensity to achieve competitive advantage in technologically leading-edge and growing activities and sectors. Innovation is now recognised as a key ingredient and as one of the main prerequisites of sustainable development of regions, nations, sectors and firms.

There are different conceptualizations on innovation systems: National Innovation Systems (NISs) (Porter, 1990; Lundvall, 1992), Regional Innovation Systems (RIS) (Cooke et al., 1997), Sectoral Innovation Systems (Breschi and Malebra, 1997) and Technological Systems (Carlsson, 1995). The importance of Regional Innovative Systems (RIS) in the economic development of regions has been posited by a number of authors (Niosi, 2008; Cooke, 2007; Vigier, 2007, Ketikidis et al., 2010) and examples of regional excellence have been identified around the world (Pellegrin, 2007; Freeman, 2002). The concept of Regional Innovation System (RIS) has been the central goal of the European technology and innovation policy. This concept is considered to contribute to the Lisbon strategy by enhancing European regional competitiveness (RC) (Bruijn and Lagednik, 2005). The literature suggests that RIS possesses two sides: the supply side and the demand side (Cooke, 2007; Laurentis, 2006; Harmaakorpi, 2006). The supply side includes institutional sources of knowledge generation and institutions accountable for the preparation of qualified labour (Cooke, 2007). The demand side incorporates the productive systems, companies that apply the scientific output of the supply side (Laurentis, 2006).

The main objective of this paper is to identify and assess knowledge assets and flows, as well as interactions of SMEs with support organizations, in Northern Greece, and accordingly provide policy recommendations for the future economic development in the region. The paper is structured as follows. The following section briefly outlines the important role of innovation and regional innovation policies as a driving force of economic development in knowledge-driven economies. This is followed by a conceptualisation of the competitiveness of Central Macedonia (CM) and East Macedonia-Thrace (EMTH) regions, set within the context of existing available indicators. Finally, the final two sections focus on the methodology used, an interpretation of the data gathered and generated, resulting in some concluding remarks concerning policy interventions.

## 2 Innovation in the knowledge-driven economy

Early work by Thurow (1993) determined that it is primarily knowledge-based industries within which a nation needs to specialise in order to obtain a word-class standard of living for its citizens.

Knowledge-driven economy is "one in which the generation and the exploitation of knowledge has come to play the predominant part in the creation of wealth. It is not simply about pushing back the frontiers of knowledge; it is also about the more effective use and exploitation of all types of knowledge in all manner of economic activity" (p. 53) (EC, 2003)

According to EC (2004), the major transformations in the current economies which have been brought by the knowledge driver are the following:

- Breakthroughs in information and communication technologies have decreased costs of knowledge activities, for instance, costs of knowledge transfer;
- Knowledge is becoming a commodity. It is produced, bought and sold to a degree that has been never seen before;
- Interconnectivity among knowledge agents has deployed substantially (p.35).

The emergence of knowledge-based economy, where knowledge is the basic strategic resource to achieve growth, defines the new era in development and growth. Thus, economic development strongly depends on the country's ability to foster an environment that supports innovations and usage of new technologies in knowledge driven economy (KDE) (Niosi, 2008; Asheim, 2007). Globalisation is a difficult challenge for all regions which also offers developmental opportunities for those capable for designing appropriate policies to grasp them.

## 2.1 Regional innovation policy and policy implications

Komninos and Tsamis (2008) state that from a policy perspective understanding innovation as a systemic process has important implications for policy-makers and for identifying effective innovation policy measures. There is an increasing interest among regional state authorities to the Regional Competitiveness (RC) and Regional Innovative Capability (RIC) of their regions linked to the local firms' capabilities to innovate (Cook and Memedovic, 2003). One of the major functions of the policy-maker is to decrease uncertainty by providing information, to govern conflicts and to give incentives. These may be determined as the "universal" mission of policy-maker (Edquist, 2000). Regional Innovation Policies (RIP) are supposed to consider all economic players of Regional Innovation Systems (RIS) (Sanz-Mendez and Cruz-Castro, 2005).

According to Andersson and Karlsson (2004), in order to enhance RIS, RIP should include such measures as:

- Develop regional knowledge providers and link the companies to external knowledge sources;
- Attract skilled labour and promote the education of labour;
- Develop an institution responsible for scanning markets and technologies for regionally important clusters;
- Promote interaction and collaboration between firms, knowledge institution and governmental institutions;
- Promote recurrent contact between businessmen, developed more formal and planned networking;
- Secure the supply of venture capital (p. 78).

# 3 Description of the Thessaloniki metropolitan region

The region of CM dominates the regional economic outlook in all respects mostly due to Thessaloniki and its key role as a regional economic communications and research and development hub. Thessaloniki is the second largest city in Greece with population over one million in its metropolitan region. Unlike much of the northern Greece, Thessaloniki metropolitan region has an economic structure dominated by manufacturing and services rather than agriculture. On the other hand the region of EMTH is far less developed despite significant progress that has been taken place over the last decade, particularly in Thrace.

## 3.1 Competitiveness

The competitiveness of the economy of CM and EMTH regions lags behind the national average. The high unemployment rates are a key issue that at least for the region of EMTH is related to low productivity and the same applies for most parts of CM, if Thessaloniki was to be excluded. The weak economic structure, linked to traditional sectors with low knowledge intensity and less efficient human and physical capital deployment are the key factors. The key competitiveness policy tool for CM and EMTH are the Regional Operating Plans (ROP), implemented via the respective regional authorities. The ROPs are the tools for regional economic planning that lay out the strategies and for each region over a six- year period (currently 2007-2013). The key stakeholders involved in policies relating to Research and Development (R&D) investment are the National Government (ministries of Economy and Development and General Secretariat for Research and Technology), the Regional Authorities of CM and EMTH as well as the General Secretariat of Macedonia and Thrace.

## 3.2 Innovation and R&D

Greece and its innovation system placed at the very bottom for all 13 Greek regions (NUTS2) of the relative rankings of 203 regions with only the capital of Attica having a relatively good performance (86th) (Hollanders, 2006). Business sector expenditures on R&D are continuously less than 30% of total R&D expenditures, at around 15% of the respective EU average (EUROSTAT, 2005). According to the innovation survey (CIS-4), among the firms with some form of innovative activity, around half engaged in intramural R&D and 30% on a continuous basis (EUROSTAT, 2007). This poor performance of the Greek innovation system is replicated at regional and sectoral level.The area of Northern Greece shows a low share of the private sector as a financier of research technology and development (RTD) and a very low share of financing that flows from the private sector to public RTD performers (less than 7% of private funds) and vice-versa. On the other hand, the public sector predominates with almost 70% of financing if higher education and public companies are included. Private companies also receive only 3,8% of their RTD financing from the public sector. In the services sector there are 3,2 RTD employees per 100 while in manufacturing only 1,5 making the latter responsible for the country's lower ranking.

The main weakness of the region lies in the very low level of R&D employment by the corporate sector, which is very small both in relation to the national average and the EU average. The marked exception is R&D employment in the higher education sector, which for the case of CM is even above the EU average (1,6 vs. 1,4 per million inhabitants). According to Komninos and Tsamis (2008) public sector and especially universities have maintained a rather negative attitude towards direct cooperation and funding of R&D activities by the private sector. They focus on basic research activity that in many cases does not fit with the needs of the market.

# 4 Analysis and discussion of key findings

## 4.1 Methodology

The guiding theoretical and conceptual framework was the triple helix model, which draws on relationships and interaction between academic institutions, governmental agencies and organizations, and the industry, including knowledge transfer intermediary organisations. For this purpose a triangulation method was used to integrate information on several economic drives from past empirical research, analysis of existing statistical data for SMEs in the target geographical area, as well as interviews with representatives from Universities, research laboratories, research centres, intermediaries and policy makers. The economic drives of interest included competitiveness, innovation and research and development (R&D); knowledge demand and supply; knowledge flows and networks; academia networks; financing of the knowledge economy; and future policy approaches. Following the integration of existing data on these drivers, the strengths and weaknesses were identified, and policy recommendations for future development were made in relation to the key components of the triple helix model. All the information collected was aggregated on a technical report (MIRIAD, 2008).

## 4.2 Knowledge transfer and flow

Overall, there is currently an evidence gap of the type of transferring occurring, its density and frequency, as well the flow directions. A key finding as developed in MIRIAD (2008) is that the triple-helix interaction is reliant on a group of intermediaries that are not very well connected or interfaced. From the whole conceptualisation the following is apparent for the regions of CM and EMTH considered together:

- Government – key policymakers are the GSRT within the Ministry of Development at national level and the Regional Authorities of CM and EMTH at regional.
- Business – SME dominated regional economy with low levels of R&D investment and knowledge commercialisation.
- Higher Education – large-scale knowledge creation appears mainly restricted to a small number of higher education institutions. The cases of the Thessaloniki Technology Park (TTP) / The Centre for Research & Technology Hellas (CERTH) and Urban and Regional Innovation Research Unit (URENIO) show that there is significant potential for intermediation in synergy with Higher Education, provided that the proper institutional arrangements were in place.
- Government-Higher Education Interface - key linkages between higher education and government either lack coordination and decisive initiative or are limited at both national and regional level.
- Higher Education-Business Interaction– little evidence of direct knowledge transfer and even less of reverse transfer. Intermediaries act as key facilitators, with the government and EC being the key funder. Even though, there has been a rapid increase in the number of knowledge transfer intermediaries the process is at a relatively early stage. There is a lack of intermediaries servicing higher education institutes with the main focus being instead on the corporate sector.

- Government-Business Interaction - connections between government and the business community with regards to R&D are limited at both national and regional levels or lack of decisive initiatives.

## 4.3 Summary of recommendations

The question that has dominated our engagement with various experts and institutions was the following: Does the critical-mass present in terms of physical and human resources and infrastructure include entrepreneurial acumen/culture and financial resources adequate to drive the RTD outlook to a higher level? Hence, the priority issue seems to be the creation of a critical-mass for RTD (stock of know-how) in terms of infrastructure, engagement of capable skills, entrepreneurial spirit and imaginative policy leadership that will put into use the combined resources effectively making innovation a self-sustaining process that will allow for its faster evolution.

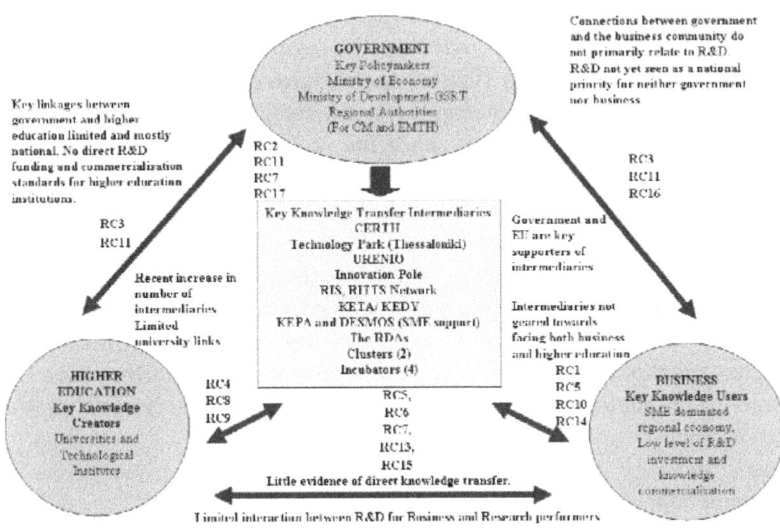


**Figure 1**: Triple helix flow model: Central Macedonia and East Macedonia-Thrace

In Figure 1, the triple helix model of the region of Northern Greece is presented along with the policy recommendations aiming to improve weak linkages of the regional innovation system. Each recommendation aims to strengthen weak relative linkages of the regional triple helix model. High priority has been given to the weakest link among higher education R&D and the private sector.

***Recommendation 1 – [RC1]*** ***Rationale***

Identify new areas in between traditional sectors where innovation can flourish, capitalizing in new technologies and shifting to new activities. Raising the interest of traditional sectors would be a challenge though, since it also relates to RTD and innovation awareness and culture. The low productivity and performance of traditional sectors of the region prompt for the identification of areas where traditional sectors could expand their activities, capitalizing on new technological advancements. For example, traditional sectors such as construction could cooperate with the automation high tech sector to produce "smart house" products.

***Recommendation 2 - [RC2]*** ***Rationale***

This recommendation applies mainly to the region of EMTH: Capitalize on the Science and Technology Legislature to improve R&D infrastructure by creating a science and technology park and relative institutes related to the R&D locally advanced areas, since the region presents high scientific skills, e.g. Biotechnology. The regional policy should be directed in creating a critical mass of R&D resources in selected research areas. The region of Northern Greece, even though it is not highly rated in R&D performance presents a high concentration of R&D infrastructures around the metropolitan area of Thessaloniki. The region of EMTH must also follow in infrastructures to equalize inequalities within the area of Northern Greece and also to capitalize on the excellences that it presents, such as the research results of the Biotechnology Department of The University Hospital of Alexandroupoli.

***Recommendation 3 - [RC3]*** ***Rationale***

Create an interregional R&D investment committee for Northern Greece, including R&D actors from Institutes, Universities, Public and Private organizations. This committee should be linked to the National Committee for Research and Technology. It must highlight the importance of also including representatives from the private organizations and intermediaries, than just academics and researchers. The strengths of the regions' favourable environment, adequate supply of scientific personnel, existence of regional funds, increasing attitude toward R&D investment by the private sector must be coordinated and organized from a widely participated body of R&D actors and business organizations, aiming to minimize the gap between R&D and the business world.

***Recommendation 4 - [RC4]*** ***Rationale***
Promote cultural awareness of innovation and entrepreneurship, especially within the community of the young generation of scientists and the future R&D human capital including high school students. This culture enhancement should be directed toward entrepreneurship and innovation.

The high level of unemployment among young scientists indicates the low level of utilization of human scientific resources within the regional innovation system. The uncertainty of the financial environment directs them to seek secure employment in the public sector. Young scientists must capitalize on the benefits and the hospitable environment of the region to create innovative start-ups. Also younger generations, such as high school students must be infused with the benefits of entrepreneurial skills.

***Recommendation 5 - [RC5]*** ***Rationale***
Increase R&D investment by private organizations outside the higher education system, by building infrastructures and providing incentives so that large scale business organizations move their R&D activities into the region. The region must implement investment policies, marketing techniques, human capital and infrastructure availability, including financial incentives so large scale corporations will benefit from relocating their R&D departments in the region. The low performance of the region in private R&D spending and patent creation indicates the inefficiency of the higher education driven R&D system. Furthermore, the creation of the area of critical mass of private sector R&D activities points towards the potential for regional growth and increase in the investment basis of the local economy.

***Recommendation 6 - [RC6]*** ***Rationale***
Capitalize on the fluent scientific resources of the higher education system by promoting synergies in transfer of know-how with private organizations. Regional incentives should be created for the transfer of R&D personnel

from the universities to industry and vice versa. The high level of scientific knowledge and human resources concentrated within the limits of the higher education system must flow outside the borders of the universities enriching the start-up repository for new business ideas and technological innovation in the area.

***Recommendation 7 - [RC7]*** ***Rationale***

Develop new regional metrics that incorporate a holistic means of measuring investment in knowledge. These indicators should measure the qualitative, "fact finding", aspects of the regional R&D performance by utilising surveys inside the regional organizations instead of statistical approaches.

Whilst R&D expenditure continues to be an important measure of innovation and the conversion to a knowledge-based economy, it is limited due to its relative inapplicability to growing service-based sectors.

***Recommendation 8 - [RC8]*** ***Rationale***

Enhance the knowledge economy by building science parks, where higher education R&D is bound to find applications and meet business demand for knowledge. The isolated higher education R&D system and the unfavourable environment for innovation commercialisation within the universities must find a hospitable environment to experiment the application of these results to the market needs.

***Recommendation 9 - [RC9]*** ***Rationale***

Enhance life long learning activities within the social structures of the region upgrading in a continuous basis the human capacity of the region. Existing quality of training curricula and services and structures should be adopted to serve long term learning action plans instead of partial vocational programmes. The high level of unemployment among unskilled workers requires immediate actions enriching the human capacity skills of the region.

***Recommendation 10 - [RC10]*** ***Rationale***

Enhance clustering support policies related to R&D activities among R&D researchers and targeted business sectors, strengthening the role of intermediaries in the cluster building process. The clustering efforts in the region and generally in the country were not adequately supported by a methodological approach that could bring together different sectors and activities into joined R&D activities. The support policies and the role of intermediaries are important elements missing from these unsuccessful clustering measures.

***Recommendation 11 - [RC11]*** ***Rationale***

Implement awareness raising policies and support mechanisms for the organizations in the private and public sector on the importance of Intellectual Property Rights (IPRs) in securing financial stability and future prospective, strengthening their R&D orientation, focusing on universities and R&D centres. The extremely low performance of the region regarding IPR management indicates the low level of R&D activities. Enhancing the value of IPRs by raising awareness and providing supporting mechanisms for IPRs, is an effective policy to increase the R&D capacity of the region.

***Recommendation 12 - [RC12]*** ***Rationale***

Implement policies to increase the inwards information flows from regional researchers and R&D institutes to local innovation actors. Although innovation actors have a profound need for knowledge creation, they rely mostly on sources external to the region. This fact weakens the opportunity to transform knowledge into regional R&D activities.

***Recommendation 13 - [RC13]*** ***Rationale***

Increase and highlight regional R&D corporate demand bringing together private organizations and the R&D actors. Policies must include proactive measures inside the business organizations to identify and match R&D demand to the regional supply. Such policies could include post-doctoral internships and incentives for building internal R&D departments. The low level of interaction among R&D organizations and the business world requires the fulfilment of demand driven R&D processes. The demand stimulates R&D performers to target their activities into applicable and commercial areas.

***Recommendation 14 - [RC14]*** ***Rationale***

Improve commercialisation of R&D results setting marketing support mechanisms and consulting services to R&D performers. The effort towards the commercialisation of R&D results is a long-term attitude changing process and is problematic within the region and countrywide. This attitude is not assisted by the current environment and commercialisation structures of the higher education systems, which fosters most of the R&D activities. Synergies must be achieved with capable marketing support organizations to promote R&D commercialisation.

***Recommendation 15 - [RC15]*** ***Rationale***
Initiate a dialogue for IPRs ownership within the region clarifying the question of "who is the proprietor?" in case of collaborative projects between Universities and the private sector. A major obstacle in the R&D collaboration efforts is the unclear status of IPR ownership within the higher education system. Private organizations are reluctant to engage in collaborative R&D projects and invest in a not well-defined and complex IPR management status.

***Recommendation 16 - [RC16]*** ***Rationale***
Identify the main corporate actors with R&D potentials and involve them into cooperative activities to increase the regional level of R&D spending

Since the business environment of the region is dominated by small and medium size businesses, the few corporate leaders must be utilised to raise R&D capacity. Smaller businesses could cluster around these corporate leaders.

***Recommendation 17- [RC17]*** ***Rationale***
Initiate a foresight exercise with a wide network of participants including intermediaries, to achieve a common understanding, as far as, the long terms aims of R&D investment regarding technologies, sectors and clustering prospectives. The low level of interaction of the regional innovation system creates a diversified view of the directions that R&D efforts and resources should take. The divergence of interests that each actor has within the system should be eliminated and efforts should be directed to common long term objectives set by the foresight exercise.

# 5 Conclusions

In the past, knowledge and R&D investment policies and strategies have focused either on stimulating transfers/spill-overs or on facilitating knowledge absorption. However, it is clear that successful strategy building must take account of both simultaneously. The aim of this study was to integrate both these aspects, so as to remove the supply and demand-side barriers associated with knowledge and R&D transfer, absorption, and investment. In essence, knowledge and R&D investment is a function of a region's ability to transfer, spill-over and absorb knowledge. In general, policy makers, in addition to specific policy targets regarding R&D spending, are aiming to stimulate demand for RTD from the corporate sector of the economy by improving its sectoral allocation, stimulating supply of RTD through the reorganization of the tertiary education and research system, opening up to global research networks and building domestic R&D capacity infrastructure.

In terms of future policy directions, it appears that research performers believe that 'Creating better networks that link companies with universities and other R&D performing organisations' together with 'Making more R&D finance available to companies enabling them to become involved further in R&D and knowledge related activities' should form the core policy issues. Significant importance is also attached to the creation of start up companies, attraction of high value foreign investment and an improved system of business support and advice. This result shows the increasing awareness of research performers in the need to address corporate requirements through stronger links between companies and R&D performing organizations (MIRIAD, 2008).

This study has capitalized on all available statistical data sources in the analytical process and validated the qualitative analysis results using an interview process with main regional actors. Since, statistical data sources had covered different chronological periods and had significant differences in the analytical approaches used, the derived strengths and weaknesses concluded from the statistical analysis are aligned with the conclusions of the regarding strengths, weaknesses and policy directions. The repetition of statistical performance rating does not support adequately policy makers towards measures required at systemic level to strengthen system performance. Although the approach to integrate specific policy recommendations to the triple helix links is a valuable policy analysis outcome, the research does not identify specific system faults.

Future directions of research must take the systemic analysis into account by using models like the Interaction, Intension, and Indicator (3I) analytical framework (Zygiaris, 2009) to relate the derived policy recommendations to specific innovation system faults. Thus, policy makers could be provided with necessary systemic performance rating to address measures to related system actors aiming to improve system performance. In turn, the derived policy recommendations must be analysed further to the extent of applying them at regional system level. This analytical process must be redefined in terms of the actors involved in each recommendation and the measures need to be taken at the system level.

## Acknowledgements

The authors are grateful to South East European Research Centre researchers that participated to the project as well as to the European Commission's Regions of Knowledge 2 initiatives for funding the project entitled 'Managing and Infusing Research Investment and Development (MIRIAD). Another word of gratitude to all project partners that contributed to the MIRIAD project (www.miriad.org).

## References

Andersson, M. and Karlsson, C. (2004) "Regional Innovation Systems in Small & Medium-Sized Regions", Innovation, Vol 15, No. 2, pp 102-132.

Asheim, B. (2007) "Differentiated Knowledge Bases and Varieties of Regional Innovation Systems", Innovation: The European Journal of Social Sciences, Vol 20, No. 3, pp 223-237.

Cooke, P. (2007) "To Construct a Regional Advantage from Innovation Systems First Build Policy Platforms", European Planning Studies, Vol 15, No. 2, pp 179-194.

Arundel, A., Lorenz, E., Lundval, B. and Valeyere, A. (2007) "How Europe's Economies Learn: a Comparison of Work Organisation and Innovation Mode for the EU-15", Industrial and Corporate Change, Vol 16, No. 6, pp 1175-1210.

Breschi, S. and Malebra, F. (1997) Sectoral Systems of Innovation: Technological Regimes, Schumpeterian Dynamics and Spatial Boundaries, in C. Edquist (Ed). Systems of Innovation, London: Printer.

Bruijn, P. and Lagednik, A. (2005) "Regional Innovation Systems in the Lisbon strategy", European Planning Studies, Vol 13, No. 8, pp 1154-1171.

Carlsson, B. (1995) Technological Systems and Economic Performance – the Case of Factory Automation, London:Kluwer.

Cook, P. and Memedovic, O. (2003) "Strategies for Regional Innovation Systems: Learning Transfer and Applications", Policy Papers of The United Nations Industrial Development Organization, [online], http://www.unido.org/fileadmin/user_media/Publications/Pub_free/Strategies_for_regional_innovation_systems.pdf

Cooke, P., Gomez Uranga, M. and Extebarria, G. (1997) "Regional Systems of Innovation: Institutional and Organisational Dimensions", Research Policy, Vol. 26, pp. 474-491.

Edquist, C. (2000) "The Systems of Innovation Approach and Innovation Policy: an Account of the State of the Art", Systems of Innovation: Growth, Competitiveness and employment, Vol 21, No. 2, pp 54-76.

European Commission (2003) "Growth, Competitiveness and Employment: the Challenges and Ways Forward into the 21st century – white paper", COM(93) 700 final. Brussels. [Accessed on the 10th March 2010].

European Commission, Information Society Technologies (2004) "Work program for 2005-2006: a Thematic Priority for Research and Development under the Specific Program - 'Integrating and Strengthening the European Research Area'", [Accessed on the 10th March 2010].

EUROSTAT (2005) "Science and Technology Statistics", EUROSTAT, [online], http:/epp.eurostat.cec.eu.int/portal/page?_pageid=0,1136250,0_45572552&_dad=portal&_schema=PORTAL

EUROSTAT (2007) "Science and Technology Statistics", EUROSTAT, [online], http:/epp.eurostat.cec.eu.int/portal/page?_pageid=0,1136250,0_45572552&_dad=portal&_schema=PORTAL

Freeman, C. (2002) "Continental, National and Sub-National Innovation Systems-Complementarity and Economic Growth", Research Policy, Vo 31, No. 1, pp 191-211.

GSRT (2004). "National Innovation Survey (CIS III) in Greek Enterprises 1998-2000", Athens: GSRT - General Secretariat for Research and Technology - Ministry of Development.

Harmaakorpi, V. (2006) "Regional development platform method (RDPM) as a Tool for Regional Innovation Policy", European Planning Studies, Vol 14, No. 8, pp 1085-1104.

Hollanders, H. (2006) "2006 European Regional Innovation Scoreboard", European Commission - Directorate General Enterprise.

Hollanders, H. and Arundel, A. (2005) "European Sector Innovation Scoreboards", European Commission - Directorate General Enterprise, European Trendchart on Innovation.

Ketikidis, P.H., Miroshnychenko, I. and Zygiaris S. (2010) "A Proposed Framework of Regional Innovation System: the Case of the Kharkiv Region in Eastern Ukraine", Paper read at 3rd International Conference on Entrepreneurship, Innovation and Regional Development, Novi Sad, Serbia, May.

Komninos, N. and Tsamis, A. (2008) "The System of Innovation in Greece: Structural Assymetries and Policy Failure", International Journal of Innovation and Regional Development, Vol 1, No. 1, pp 1-23.

Laurentis, C. (2006) "Regional Innovation Systems and the Labour Market: A Comparison of Five Regions", European Planning Studies, Vol 14, No. 8, pp 1059-1084.

Lundvall, A.B. (1992), National Systems of Innovation: Towards a Theory of Innovation and Interactive Learning, London, Pinter.

Lundvall, B. (2007) "National Innovation Systems-Analytical Concept and Development Tool", Industry and Innovation, Vol. 14, No. 1, pp 95-119.

MIRIAD (2008), "Managing and Infusing Research Investment and Development: Final Report - Central, East Macedonia and Thrace knowledge Investment Strategy", South East European Research Centre, Thessaloniki- Greece.

Niosi, J. (2008) "Technology, Development and Innovation Systems: an Introduction", Journal of Development Studies, Vol 44, No. 5, pp 613-621.

Pellegrin, J. (2007) "Regional Innovation Strategies in the EU or Regionalised EU Innovation Strategy? " Innovation, Vol 20, No. 3, pp 203-221.

Porter, M. (1990) The Competitive Advantage of Nations, New York: The Free Press.

Sanz-Mendez, L. and Cruz-Castro, L. (2005) "Explaining the Science and Technology Policies of Regional Governments", Regional Studies, Vol 69, No. 2, pp 345-370.

Storper, M. (1997) The Regional World: Territorial Development in a Global Economy, New York: Guilford Press.

Thurow, L. (2003) Head to Head: The Coming Economic Battle Among Japan, Europe and America, London, Nicholas Brealey.

Vigier, P. (2007) "Towards a Citizen-Driven Innovation System in Europe", Innovation, Vol 20, No. 3, pp 191-202.

Zygiaris S. (2009), "Envisaging Regional Innovation System Failures and Highlights", Romanian Journal of Regional Science, Vol 3. No. 2, pp 2-5.

# The Steps of an Organisational Routine's Transformation:
## The Case of the Employee Driven Innovation Routine at the French National Railways Company (SNCF)

**Carine Deslee**
Université de Lille 2, France

**Editorial commentary**

Deslee explores the notion of the 'Participative Innovation Routine' (employee-driven innovation) in the context of the French National Railways Company (SNCF) at a time when the organisation is faced with the dual challenges of markets opening and competition increasing on an international level. In a study conducted over a period of many years, the author investigates how the Participative Innovation Routine has changed over time.

A range of approaches was used to gather data to inform this case, including internal organisational documentation, archival documentation, interviews, direct observation, participant observation, and the analysis of physical and cultural artefacts.

Some points for learning and discussion arising from this case include:

- The changing nature of the Participative Innovation Routine.
- The role of employee reward systems in fostering innovation.
- The management of employee involvement in innovation.

- The pros and cons of centralised versus de-centralised approaches to employee-driven innovation.
- The barriers to innovation in large, bureaucratic organisations.

---

**Abstract**: In a context of competition, innovation is a source of enterprise development. Our research focuses on the SNCF company with 160.000 railworkers. It is a bureaucratic structure (Weber 1971, Mintzberg 1982 and Alter 1995, 1996). The SNCF activities concern various sectors such as freight, equipment, infrastructure and travelers in France and Europe. The company faces a large number of challenges in this context of openness to competition. We analyze the Participative Innovation routine that has deeply changed since the 1990s although a routine in the literature is considered to be stable (Nelson and Winter, 1982 ; Cohen and al., 1996). Our questions include how this Participative Innovation routine has changed so deeply ? How could a suggestion system limited to the field maintenance become one of the pillars of the SNCF's strategic plan, called Industrial Project ? Since 2005 this routine has become a tool of communication both internally and externally as part of the campaign "Ideas ahead" ("des idées d'avance"), which became a part of the SNCF company's identity. We rely on the methodological approach by Barley and Tolbert (1997) since 1990 to understand the different steps of an organisational routine's transformation through the notion of script. We adopt an interpretativist posture (Girod-Seville and Perret, 1999) to study this Participative Innovation routine understood in the sense of Feldman (2000). We show that this routine has changed and we describe its interrelations with the SNCF company. The interest of this research concerns the railways companies but more generally all companies involved in market opening and competition. Under the impetus of the strategic project of the SNCF company, we show why and how this Participative Innovation routine has changed over the years.

**Keywords**: routine, participative innovation, innovation, SNCF, script

# 1 Introduction

Faced with international competition, innovation is a key to business development. We consider here the case of the SNCF company as a bureaucracy and more particularly to the routine of Participative Innovation, which is an employee driven innovation routine. Its aim is to involve all stakeholders of the company in search of innovations, both in terms of products and services. Our interest in this research is how the SNCF company can successfully meet the challenges ahead, including the opening of its market and competition.

The corporate branch of the SNCF company has made innovation a participative tool in its human resources policy which is included in the strategic plan of the company. It uses it as a vehicle for communication both externally and internally since april 2005, which contributes to change the corporate culture. We have studied the evolution of the Participative Innovation routine for over twenty years. Our question is as follows: how an organizational routine such as Participative Innovation has transformed over time? How has it turned from a limited suggestion system when the SNCF company was created in 1938to become one of the levers of the company's strategic plan, called industrial project? After clarifying our theoretical framework and the concept of routine, we present the Participative Innovation routine in the strategic context of the SNCF company and the design of our research before a final section on analysis and results.

# 2 Theoretical framework of our research

## 2.1 The concept of routine

There is no unified vision of the concept of routine. Authors have very different meanings depending on whether it includes a model of behavior in individual or organizational level or learning ability. Evolutionary theory provides a special place to routines. The routines are defined by Nelson and Winter (1982) in this context as "behavior patterns regular and predictable". The routines are tacit in nature and stable (Autissier and Wacheux 2000: 40). Cohen and al. in August 1995 (from Feldman 2000 : 611-612 and Reynaud 2001) attempted to clarify the concept of routine. They define the routine as an executable capability for repeated performance in some contexts that have been learned by the organization in response to certain pressures. It is an "ability to perform an action repeatedly in a context that has been learned by an organization, this action is defined as a pressure towards the selection". The consensus of this working group did specify the routine as a model of action. The fact that an organizational routine can be transformed over time has been little studied. It is indeed generally considered as stable.

We consider this definition of Feldman (2000 : 611) to define the Participative Innovation routine in the SNCF company : "The routines are temporal structures which are often used as a way to perform organizational work." We inscribe our research in this constructivist approach with works like those of Barley (1986) or Pentland and Rueter (1994) who deal with the interactions between actors, recognized and legitimized interactions, Tsoukas and Chia (2002) or Feldman and Pentland (2003). Feldman (2000) distinguishes the 'ostensive' aspect of a routine as a standard or norm, representation of the routine (for instance a hiring routine) from the 'performative' aspect of a routine, that is to say how it is implemented. Feldman and Pentland (2003: 97) have several metaphors of organizational routine, especially one that connects to the performance programs or scripts that deal with complex social and organizational behavior (Gioia and Poole, 1984).

## 2.2 Some points on participative innovation and innovation

Let's define what we mean by Participative Innovation. Others before us have been interested in this subject. Durieux (2000) defines Participative Innovation as proposed innovation by actors of the company different from their missions. It concerns employee driven innovation. More comprehensive than the suggestion, this type of innovation involves active participation of the innovator throughout the development process. Everaere (1996) was also interested in Participative Innovation, which "emanates from a desire at the head of the company not to book innovation to a few specialists, but to spread it among all company employees". We complete this definition by Van de Ven's definition of innovation (1986: 591-592) for whom "an innovation is first of a new idea. The idea of 'new' can be a recombination of old ideas as long as the idea is perceived as new by the individuals involved. It is an innovation even if it may appear to others as an "imitation" of what may exist elsewhere. The innovation includes the development and implementation of new ideas by individuals who engage over time in transactions with others within an institutional order."

Methods to encourage the rise of ideas are varied. The theory of the "Idea Management" (De Bradandère, 2002) identifies four potential sources: customers, suppliers, employees and shareholders. Participative Innovation focuses on the "employees" source to develop the creative potential of individuals in an organization. Creativity is not due to chance. According to this typology developed by Slappendel (1996), we consider that Participative Innovation relays on the interactive process. We may wonder how an organization like the SNCF company is conducive to Participative Innovation. Of course, we are aware of the variance in such a large company.

## 2.3 The SNCF company, a bureaucratic structure a priori unfavorable to innovation

Innovation in established organizations is a difficult process (Weber, 1971; Mintzberg, 1982 Alter, 1995, 1996). For Weber (1971), bureaucracy is an instrument of rationalization in organizations. He described the ideal type of bureaucracy with three essential features: the impersonality of the rules, the character of expert, a hierarchical system involving mandatory subordination and control. Thus, there is potential for innovation in public services to be exploited (Clergeau de Mascureau, 1995). By nature, innovation is contrary to the essence of bureaucratic organizations (Clergeau de Mascureau, 1995: 145, 147) where any action is subject to predefined procedures. Bureaucracy is legitimate because it is effective. The bureaucracy is explicitly designed to obey and not to imagine.

Burns and Stalker (1961) introduced the distinction between the mechanistic (or bureaucratic) structure and organic structure. The first structure is considered quite suitable for the performance of standardized tasks in a stable environmental context, while the second one applies to situations marked by technological or commercial instability. The organization of the organic type is more likely to innovate because it shows more adaptability in changing environments than the mechanistic type. A number of authors (Burns and Stalker, 1961; Zaltman et al., 1973) underscore the need for a less formal, less bureaucratic structure to promote independence and creativity for the actors (Burns and Stalker, 1961).

Normann (1971: 214) is joined in his analysis by Adler and Boris (1996) which specify that the bureaucracy can be facilitative or coercitive as appropriate. Other authors, such as Spender and Kessler (1995: 43), argue that innovation requires at every stage of development a mode of organization that is more favorable.

The SNCF company can be considered as a mechanistic bureaucracy according to Mintzberg's typology (1982) and a mechanistic organization by Burns and Stalker (1961).

In a bureaucratic context such as described for the SNCF company, the existence of a Participative Innovation routine is interesting. Bureaucracy seems not to be the most favorable to innovation even if there are margins of freedom for the actors as confirmed by Bouyer et al. (2003) in their study on the railways culture. Our question is how an organizational routine, such as the employee driven innovation routine, evolves over time and takes into account the challenges facing the SNCF company.

## 2.4 Research question and overview of the theoretical framework proposed by Barley and Tolbert (1997)

Our purpose is to understand why and how an organizational routine such as the employee driven innovation routine can transform itself and accompany the evolution of an organization in the case of the SNCF company.

We rely on the simplified sequential model of institutionalization proposed by Barley and Tolbert (1997) applied to the case of the SNCF company to highlight the different scripts and Participative Innovation routine in this context. An institution consists of cognitive, normative and regulative structures that bring stability and give meaning to social behavior (Scott 1985 : 33). Institutions are conveyed in different ways (culture, structures and routines) and intervene at different levels. Barley and Tolbert (1997) defined the institution through shared rules and typologies that identify categories of social actors and their activities or relationships.

The theoretical framework proposed by Barley and Tolbert (1997) is based on the notion of script. These scripts can be empirically identified regardless of the type of actor or the level at which interest the researcher. According to Barley and Tolbert (1997 : 98), scripts are observable. These are recurrent activities and patterns of interaction characteristics in a particular context. Scripts can be identified empirically. This sequential model is interesting to describe the institutional realities and the realities of the action of changing the Participative Innovation routine. By institutional reality, we mean structures with routine behavior considered 'for granted' (DiMaggio and Powell, 1991; Scott, 1995). From these everyday interactions that occur in institutions, it is possible to define scripts (Barley 1986). The reality of the action corresponds to the reality of what occurs at the level of actions. We present this Participative Innovation routine in the context of the SNCF company.

# 3 Participative innovation in the SNCF company

## 3.1 The strategic context and the participative innovation routine in the SNCFcompany

The SNCF company is a public company, controlled by the state. It has a long history that began in 1938 with the merge of five private companies and two public networks. It is first a company controlled by the state. Then, in 1983, it became a public industrial and commercial Enterprise (EPIC). These measures come within the framework law on inland transport (Loti) of 30 December 1982.

This 'social body' (Bouyer and al. 2003: 52) defends the existence of the social contract (status, retirement...). The great strikes of 1986-1987 can be explained through the threaths for these foundations of the social contract. In the early 1990s, French companies such as EDF-GDF, SNCF and La Poste have undergone significant organizational changes to achieve the necessary flexibility to deal with private competition and to adapt to new EU directives. In 1995, the European Conference of Transport Ministers held an international seminar to discuss the situation of railways transport in Europe, whose results were compiled in the publication "The railways, for what ?". In discussing development opportunities for the railways in the near future, Plassard (1995) argues that the railways are in a crucial period where the services to develop and those to stop must be defined. The European directive of 29th July 1991 led to the accounting separation of infrastructure and operations and to the creation of RFF (Réseau Ferré de France) in 1997.

We present in Figure 1 the main issues in the strategic context of the SNCF company.

In 1996, one the responses of the SNCF company to the new regulations is the development of an Industrial Project seen as the strategic response by the company. Different approaches to innovation have been launched by the SNCF company in recent years. The Participative Innovation routine is not the only form of innovation launched by the SNCF company. Our research object concerns the Participative Innovation routine, so we do not develop further these other elements such as the creation of subsidiaries, research programs, centers of expertise or political engineering innovation... We can note however that the research programs, the engineering of innovation are the responsibility of the Innovation and Research Department, including new missions in 2005 such as building and piloting a comprehensive policy of innovation.

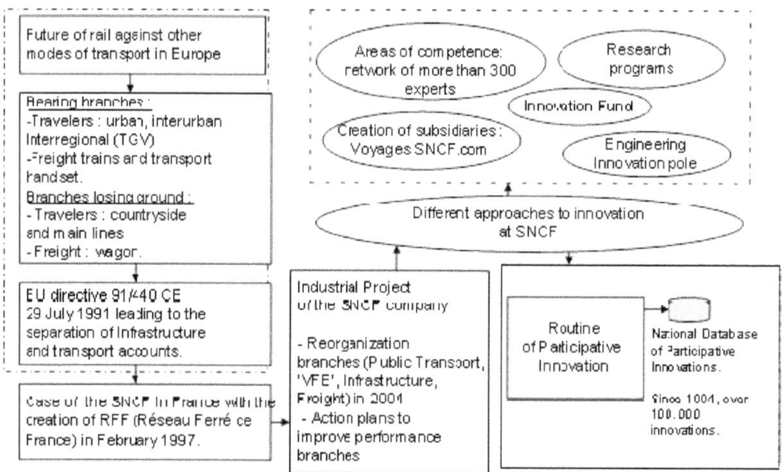


**Figure 1**: The strategic context and the participative innovation routine of the SNCF company

## 3.2 The participative innovation in the SNCF company : A routine that goes back to 1938

The Participative Innovation in the SNCF company has a long history. Its roots go back to "suggestions" that have existed since the creation of the SNCF company in 1938. In the Equipment, the creation of new tools and innovation on a daily basis were valued and included in repositories ("référentiels"). We worked on retrospective accounts and use of archival data useful even if the stories during the action for the period 2003 to 2006 are less prone to reconstruction streamlined. Thus, the suggestions "looking for all the savings in the operation" (Agenda No. 7 signed by the Managing Director A. Besnerais, May 25, 1938) are controlled by the technostructure with the possibility for any railroad workers to write directly to the Director of the technical services organization. The agenda of May 25, 1938 followed by a General Instruction, Administrative Series, series 1 General Affairs of 20 June 1939 put forward the broad principles of Suggestions, ancestor of the Participative Innovation routine with the research of savings and individual participation. The statutory of June 20, 1939 signed by the Director General Robert Besnerais plans to send suggestions to the regional headquarters (North, East, South East, Mediterranean, Southwest and West) or headquarters (branch) from Policy Statement. In 1942, a thousand of "suggestions" were produced. With the postwar reconstruction of the network that number of "suggestions" increased.

We present the analysis and results in section 5 entitled "Analysis and Results" concerning the transformation of the Participative Innovation routine since 1938 by relying on the methodological approach described in the research design that we present below.

# 4 Research design

Our approach is interpretativist (Girod-Seville and Perret, 1999). We analyse a set of contextual elements to understand events over time. Indeed, we recognize to interpret the data to ensure consistency while it is clear that we strive to maintain the neutrality necessary for research. In this single case study, we use both the documentation, archival records, interviews, direct observation, participant observation, and physical and cultural artifacts. Our approach and our ties with the field allow us to access all these data sources as part of our case study of the SNCF.

After an initial investigation in 2004, we contractualised our presence in this company by a convention of study and research in 2005. We describe the different steps of our research approach in figure 2.

In collecting and presenting data, the three principles outlined by Yin (1994) were respected.

To define and study the various scripts of Participative Innovation routine, we adapt some methodological benchmarks recommended by Barley and Tolbert (1997). Our approach includes actions until the various informations collected make sense of the scripts. We have yet to distance what has been done by Barley and Tobert (1997) to adapt their proposals to the methodological framework of our research.

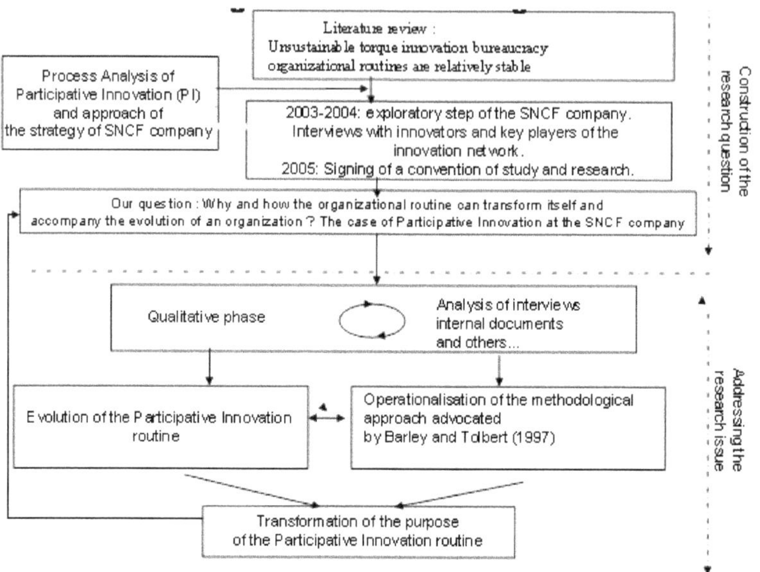


**Figure 2:** Research design

## 4.1 Our approach is summarized by the following main steps.

The first step is to first determine the scope of our study under the framework of Barley and Tolbert (1997). We must determine studied groups of social actors to analyze concerning Participative Innovation. In our research, we study the Participative Innovation routine in the SNCF company and its transformation over time since its inception in 1938. We believe that Participative Innovation is an institution in the sense that this device is "institutionalized" in the SNCF company.

As Barley and Tolbert (1997) point out, their methodology can be applied at various observation levels from microscopic to macroscopic. We chose a macroscopic level, given the size of the SNCF company. The choice of sites observed is relevant with the nature of the Participative Innovation routine which is controlled centrally. We relied on rules and procedures or internal memos issued by the national level with a national scope of application. The construction and validation of scripts required the collection and structuring of numerous data. To study a company of such a size, it is necessary to establish a database (second step).

The second step in the establishment of a database on the Participative Innovation routine is an essential step in our methodology. The number of steps that we follow is five because of this second step when Barley and Tolbert believe in four. Our presence in the company ranges from 2003 to 2007. For this period, we have compiled records of specific comments. However, it is not possible to access information on all previous periods. We used a variety of archival sources (references, notes, memos, minutes of meetings, transcripts of official speeches) but also some retrospective accounts whenever possible. Finally, through our research convention signed with the SNCF company in 2005, we had access to a database extracted from the Intranet system of the company since 1994. Thus, these databases provide evidence in terms of number of innovations produced by the Participative Innovation routine.

For the period from 2003 to 2007, we used archival records, interviews and observations. Our archival work concerns different denominations of the routine (suggestions, Innovation, Innovation, Continuous Improvement and Participative Innovation). We summarize the sources of usable information according to the periods in table 1.

**Table 1**: Sources of information on participative Innovation routine in the SNCF company according to the periods observed

| Period | 1938 to 1993 | 1994 to mid 2003 | Mid 2003 to mid 2007 |
|---|---|---|---|
| Data Archives (rules and procedures) | Yes | Yes | Yes |
| Computer databases of records Innovation | No | Yes | Yes until 2004 |
| Retrospective accounts | Difficult | Achievable | Made |
| non-participant observation | No | No | Yes |

We conducted 70 interviews with operational railworkers, 15 with experts and 23 with innovators. We structured the data to deal with them, this is the third step. The purpose is to then formulate and analyze scripts. The aim of these scripts is to better understand the change over time. For this work of structuring data before being able to exploit the wealth of information and highlight the script, Barley and Tolbert (1997) propose several ways of grouping. First, the use of rules and procedures help us to distinguish clear groups of actors, some of which appear over time. Finally, we classify the observations by identifying different types of behaviors that may occur over time. We seek to highlight periods of stability of the Participative Innovation routine which are the scripts.

This is the subject of the fourth step followed. First, we rely on the analysis of trends. Initially, our strategy is to target events that maximize the likelihood of change in Participative Innovation routine. Other endogenous phenomena and events related to the transformation process can of course lead to institutional change. It should be studied in a second time.

Each significant change in the Participative Innovation routine can be a sign of a new script, since it is by definition a stable period. It is important here to discern major changes from minor modifications of the Participative Innovation routine.

Finally, a fifth step for connecting the scripts from compliance with other sources of data on changes in the Participative Innovation routine, which led in practice to enhance our database on the Participative Innovations with new features over time. We used data relating to both a qualitative and a quantitative approach that can be complementary and articulated. This dual approach makes sense in terms of what Barley and Tolbert (1997) advocate, i.e. we use all available data.

We now present in section 5 simplified scripts that have been highlighted over the past twenty years for the Participative Innovation routine.

# 5 Analysis and results

The complexity of the Participative Innovation routine regarding the SNCF company, led us to decompose it into basic routines, which makes it easier to conduct our analysis (4.1.), Then to complete our analysis, we make an analysis concerning the scripts (4.2).

## 5.1 An analysis of the participative innovation routine of the SNCF by four basic routines

The basic routines that we have identified are the issue of an idea, the examination of an idea, the recognition and animation and management of Participative Innovation routine.

The issue of an idea is a routine that describes how an agent should proceed to submit innovative idea in the SNCF company.

The routine of examination of an idea concerns the management of the proposed idea of innovation from the deposit until a decision is taken (decision applying or not). This examination may involve multiple actors, hierarchical or functional. This is to evaluate the interest of the idea, to identify the feasibility of using experts to examine and determine the appropriate scope of application. It should also check if a similar idea was not proposed or to quantify the costs and potential earnings.

The routine of recognition focuses on how to manage the policy of rewards. This recognition routine is triggered when the statement closes the file. This recognition can take various forms such as the involvement of the agent innovator in the development of the idea, participation in events or the award of a bonus.

The animation and management routine determines how to animate the Participative Innovation routine. This covers aspects of statistical monitoring filed and processed, but also communication about the approach of Participative Innovation routine in the SNCF company, including through the organization of events such as "challenges".

There is no specific routine monitoring the implementation of an innovation once the implementing decision is made.

For the examination of an idea, we classify information according to the management of the idea, the concept of admissibility of an idea or the criteria of classification, the attitude prescribed to the manager, the help provided to the innovator, the time required to investigate the idea.

For the recognition of an idea, we consider the policy of reward, the principles of calculating a bonus, the anticipated involvement or not of the innovator in the development of the idea and events, "challenges" scheduled for innovators.

To complete our understanding of how the Participative Innovation routine has transformed itself and accompany the evolution of the SNCF company, we will rely on an analysis of the scripts according to the method recommended and adapted from Barley and Tolbert (1996).

## 5.2 Analysis of results by the scripts

Following these reviews, we identified three potential successive main scripts that correspond to major changes determined through the collection and structuring of data.

As we focus on collectives, dating from the beginning of a script is not close to a year. These scripts are then tested potential with all the information collected and structured. The first script called routine of "bureaucratic suggestions" and goes from 1938 to 1989. It covers the period of merge between the different railways companies with the period of world war and reconstruction of the network. Then gradually, between 1954 and 1972, the rules and procedures are more precise. However they do not differentiate between the local, regional and national innovation. Innovations with national impact have to go through all hierarchical levels to complete. This leads to long delays and discouragement. From 1973, the circuits of ideas differentiate innovations that are local or regional versus national ones. Facilitators roles for innovation appear.

However, it was not until the following script from 1990 to 2004 that the routine is actually supported by a network of leaders dedicated to innovation in the company and in areas with specific budgets. In the 1990s, Participative Innovation is presented as a tool for managerial renovation. Then starting in 1996 with the establishment of the strategic plan, preferred innovations progressively correspond to the strategic plan.

A new script seems to emerge from 2005 : the routine enters into consonance with the strategy of the SNCF (Industrial Project). A new impetus is given to innovation in 2005 by the Direction of Communication with the campaign "Des idées d'avance" both internally and externally. This responds to the mandate received by the Department of Research and Innovation in 2004 to rethink the policy of innovation in the company.

In 2006, the Participative Innovation is helped by the creation of decision centers. Specific budgets are allocated to enable entities (institutions and regions) to test ideas for innovations without necessarily asking the approval of their hierarchy. This constitutes a major change in the routine. The process of Participative Innovation is controlled jointly by the Human Resources Department and the Direction of Research and Innovation, but also by the activities and departments, potential customers of these innovations. The characteristics of the three scripts are detailed in table 2 below.

**Table 2:** Summary of the different scripts in the participative innovation routine

| Titles | From 1938 to 1989 | From 1990 to 2004 | From 2005 |
|---|---|---|---|
| Script name | Bureaucratic routine of "suggestions" | Routine supported by a network of innovation animation. | A routine in consonance with the strategy of the SNCF company. |
| Terminology | Suggestions | Innovation | Participative Innovation |
| Perimeter | Saving and improving the functioning of the whole company. | - Daily Progress<br>- Strategic Challenges | - Strategic axes of Industrial Project<br>- Improvements directly applicable locally. |
| Main message | Individual Participation from 1938 to 1953.<br>- Appearance relationship from 1954 to 1972.<br>- Social climate from 1973 to 1988. | Managerial logic : you have the right to have ideas and make them up. | "Ideas in advance " since 2005.<br>"The quality of the ideas, ideas of quality." |

| Titles | From 1938 to 1989 | From 1990 to 2004 | From 2005 |
| --- | --- | --- | --- |
| Central entity controlling the routine | - Technostructure<br>- Administration Direction in charge of Human Resources from 1973. | Co-piloting the Human Resources Department and the Direction of Innovation and Research. | Co-piloting by the Human Resources Department, Direction of Research and Innovation and Operations Directorate and Estates. |
| Tools | - Register book and creation of a suggestion form in 1954,<br>- Circuit varies according to the scope of innovation : local regional, national beginning in 1973. | - File formal paper.<br>- Minitel in 1994.<br>- Computer application | Consolidation of the computer system : Innogest is replaced by the new system J'Nov. |
| Allocated resources | No specific resources. | Network of local leaders, regional and national established in 1991.<br>- Time spent by the permanent network of innovation is more important after 1997. | Budgets are allocated at the institutions and regions to test innovative ideas. |
| Organizational Changes | - Creating the suggestions in 1938.<br>- Structuring the device issuance, training, bonus in 1954. | - The facilitator of innovation will be handling the case.<br>- Establishment of local leaders (DpX).<br>- Strengthening of the network of animation. | - Apparition of Decisions Centers.<br>- Professionalization of the network. |
| Animation | Despite a gradual broadening of the scope of innovation and a formalized system, innovators are not helped (no network of innova- | - A network of facilitators is taking place. The possibility of a help managed by the Direction of Research and Innovation.<br>- The Industrial Project | - The communication is emphasized.<br>- Leaders are better trained.<br>- Announcing a treatment goal of |

| Titles | From 1938 to 1989 | From 1990 to 2004 | From 2005 |
|---|---|---|---|
| | tion animation). | enables communication strategy.<br>- The animation of the process is strengthened.<br>- Tools like the 'case of innovation' appear. | ideas in less than three months. |
| Number of cases filed Participative Innovation | The system is formalized and slow.<br>- The company has a management style, which may discourage innovators.<br>- The number of innovations in the system stagnated at about 1.000 per year. | - Despite the lack of training and tools, caseload increases : it doubles from 1989 to 1996 with 2.200 cases in 1996.<br>- Despite difficulties, such as delays in processing, the number increases to 6.000 per year in 1997 and remains at that level. | 10.000 files in 2005.<br>20.000 cases in 2006.<br>Goal of 30.000 cases in 2007. |

These changes in the Participative Innovation routine are partly due to exogenous shocks such as the strategic corporate plan which is accompanied by a tripling of the number of filings. The change in this routine seems driven by technostructure and the will of the General Direction to develop Participative Innovation, that is to say employee driven innovation. Means are gradually allocated to Participative Innovation with dedicated resources (training, documentation, computer tools in particular). Since 2006, Decisions Centers are created.

Furthermore, we determined the efficiency of the Participative Innovation routine to assess the extent to which innovations are actually produced. For this, we have built on databases of participative innovation in the SNCF company. We analyze the contemporary period from 1994 to 2004. It shows that the portfolio of above 60.000 innovations bears an important issue for SNCF company of around 70 million Euros in recurring savings expected over 10 years. As shown in table 2, we can note an increase of the number of innovations produced in proportions that can be very significant. The number of innovators, however, remains limited across the SNCF company.

To conclude, the Participative Innovation routine will be structured over time. Correspondents will be appointed, and later dedicated resources will be allocated to this routine. It is then the creation of correspondents in the regions and under the Institute of Management impulse and the Research and Innovation Department, a dedicated team to support promising projects will be dedicated. These various resources can facilitate and professionalize the process.

The identification of limitations of this research requires to summarize some specific characteristics of our research design. Despite a willingness to explore as widely as possible the transformation of the Participative Innovation routine, we do not claim to cover all aspects of the process in action. Our research concludes that the evidence for this new last script are not fully implemented actions, which would need empirical observations over a longer period.

Our question is interesting in itself because the organizational routines are often considered as stable. The Participative Innovation has been little investigated to date. Our research also contributes to existing work on innovation in a bureaucracy in general and Participative Innovation in a bureaucracy in particular. We investigate the reality of Participative Innovation including in terms of produced innovations. This method of scripts according to Barley and Tolbert (1997) has been poorly implemented. Finally, the Participative Innovation routine was studied over 70 years, which required a large collection of data that has been made possible by a consistent presence in the SNCF company. Participative Innovation is more and more developed in companies. It is therefore important for companies to better understand this system.

## References

Adler P.S. and Borys B. (1996) "Two types of Bureaucracy : Enabling and Coercitive", Administrative Science Quarterly, Vol. 41, N°1, Mars : 61-89.

Alter N. (1996) Sociologie de l'entreprise et de l'innovation, PUF.

Alter N. (1995) "Peut-on programmer l'innovation?", Revue Française de Gestion, mars-avril-mai, n°103, p.78-86.

Autissier D.and Wacheux F. (coord.), (2000) Structuration et management des organisations. Gestion de l'action et du changement dans les entreprises, L'Harmattan, Logiques de l'action.

Barley S.R. and Tolbert P. S. (1997) "Institutionalization and Structuration : Studying the Links between Action and Institution", Organization Studies, 18/1, 93-117.

Barley, S.R. (1986), "Technology as an occasion for structuring: Evidence from observations of CT scanners and the social order of radiology departments", Administrative Science Quarterly, Vol. 31, N°1, p. 78-108.

Bouyer F., Marbot E., Normand E.and Thévenet M. (2003) SNCF : la culture cheminote, Rapport commandé par la SNCF, décembre.

Burns T and Stalker G.M. (1961) The management of innovation, London : Tavistock Publications Limited.

Clergeau de Mascureau C. (1995), "Quelles entraves organisationnelles et institutionnelles à l'innovation dans les organisations bureaucratiques publiques?", Revue Politiques et Management public, juin, cahier 2, vol. 13, n°2, p. 141-171.

Cohen M.D., Burkhart R., Dosi G., Egidi M., Marengo L., Warglien M. and Winter S., (1996) "Routines and other recurring actions patterns of organizations : Contemporary research issues", Industrial and Corporate Change, 5 (3), 653-698.

De Bradandère L. (2002) Le management des idées : de la créativité à l'innovation, Editions Dunod.

DiMaggio P.J. and Powell W.W. (1983) "The iron cage revisited : institutional isomorphism and collective rationality in organizational fields", American Journal of Sociology, vol. 48, n°2, p.147-160.

Durieux F. (2000) Management de l'innovation, FNEGE, Vuibert.

Everaere C. (1996) "Des effets contrastés de l'encouragement à l'innovation pour l'innovation de service", 5è conférence de l'AIMS (Association Internationale de Management Stratégique), Lille, 21 pages.

Feldman M.S. (2000) "Organizational Routines as a source of continuous Change", Organization Science, Novembre-Décembre, vol. 11, n°6, p.611-629.

Feldman M.S. and Pentland B.T. (2003) "Reconceptualizing Organizational Routines as a Source of Flexibility and Change", Administrative Science Quarterly, 48, p. 94-118.

Gioia D. and Poole P.P. (1984), "Scripts in Organizational Behavior", Academy of Management Review, vol. 9, n°3, p. 449-459.

Girod-Séville M. and Perret V. (1999), " Fondements épistémologiques de la recherche", in Thiétart R.A. and al.., Méthodes de recherches en management, Dunod, Paris, p.13-33.

Mintzberg H. (1982) Structures et dynamiques des organisations, Editions d'Organisation, Paris.

Nelson R.R. and S.G. (1982) An evolutionary theory of economic Change, Cambridge, MA, The Belknap Press of Harvard University Press.

Normann R. (1971) "Organizational innovativeness : Product variation and reorientation", Administrative Science Quarterly, vol. 16, n°2, p. 203-215.

Pentland B.T.and Ruetler H. (1994) "Organizational Routines as Grammars of Action", Administrative Science Quarterly, vol. 39, n°3, sept., p. 484-510.

Plassard F. (1995) A une nouvelle demande, un nouveau service, in CEMT, Conférence Européenne des Ministres des Transports, Des chemins de fer, pour quoi faire ?, Séminaire International, 19-20 janvier, 99-135.

Reynaud B. (2001) " 'Suivre des règles' dans les organisations", Revue d'Economie Industrielle, N°97, 4è trimestre, 53-68.

Scott W.R. (1995) Institutions and organisations, Sage Publications.

Slappendel C. (1996) "Perspectives on Innovation in Organizations", Organization Studies, 17/1, 107-129.

Spender J-C. and Kessler E.H. (1995) "Managing the Uncertainties of Innovation : Extending Thompson (1967)", Human Relations, Vol. 48, N°1, 35-56. Tsoukas H. and Chia R. (2002) "On organizational becoming : rethinking organizational Change", Organization Science, Septembre - Octobre, Vol. 13, N° 5, 567-582.

Van de Ven A. (1986) "Central problems in the Management of Innovation : implications for integration", Management Science, Vol. 32, 590-607. Weber M. (1971) Economie et Société, Les catégories de la sociologie, tome 1, Pocket. Yin R.K. (1990) Case Study Research : Design and Methods, 6th printing, Newbury Park, CA, Sage.

Zaltman G., Duncan R. and Holbek J. (1973) Innovation and Organizations, John Wiley, New York.

# Academic Intrepreneurship: Transition Strategies for Commercialisation of High Volume Electronics Products in a South African University

**Jonathan Youngleson and S Jacobs**
Tshwane University of Technology, Pretoria, South Africa

**Editorial commentary**

Youngleson and Jacobs present a case about academic entrepreneurship, focusing on a higher education initiative to foster technological innovation through applied research and development in South Africa. They outline the various strategic strands of this initiative including opportunities for student start-ups and prototype manufacturing facilities for market testing. Their discussion is concentrated in the context of the manufacture of a product in the energy load management sector that was developed under the auspices of the technological innovation initiative.

This detailed case was constructed using action research techniques, permitting investigation of the institutional transformation that took place as well as the tracking of individual participants who were affected by that transformation.

Some points for discussion and learning arising from this case include:

- The resources and infrastructure required to transform traditional higher education establishments into entrepreneurial organisations.

- The nature of the organisational culture required to create an entrepreneurial higher education establishment.
- The similarities and differences between a traditional academic knowledge and skills set and a more entrepreneurial academic knowledge and skills set.
- The contribution universities can (or should?) make to fostering technological innovation and entrepreneurship.

---

**Abstract**: Tshwane University of Technology (TUT) was established in 2004 as a new higher education institution through the merger of three previous technikons, Technikon Pretoria, Technikon Northern Guateng, and Technikon Northwest. TUT has the Vision 'to be the leading HEI (in Africa) with an entrepreneurial ethos ...'. In support of this Vision TUT has stated its Mission as to 'extend the parameters of technological innovation by making knowledge useful through focused applied research and development'.

There are several major strategies which have been introduced by the university to bring this effort to realization, and among these are the establishment of the French' South Africa Technology Institute in Electronics (F'SATIE) and its associated in-house pre-incubator Incentif. Incentif provides an opportunity for students to establish new spin-out enterprises based on technology developed in the university.

In close proximity to F'SATIE there has been established with the support of the Department of Science and Technology a Technology Station in Electronics (TSE) of the Faculty of Engineering and the Built Environment. The TSE provides for a small-scale but sophisticated state-of-the-art electronics manufacturing capability. The concept is that products developed on campus can be built up as prototypes, and then manufactured in small volume, before being transferred to a large-scale manufacturing facility. This strategy allows markets to be tested, and early stage commercialization activities to be tested before full-scale roll-out commences.

This paper looks at the manufacturing processes and challenges of a specific detailed case study of a high technology product in the energy load management sector developed with support from the Innovation Fund by a consortium including Tshwane University of Technology.

**Keywords**: Electronics manufacturing university commercialisation

# 1 Background

Following the January, 2002 Lekgotla (high level meeting), the South African Cabinet directed the Minister of Arts, Culture, Science and Technology to produce a National Research and Development (R&D) Strategy (DACST 2002). In its R&D Strategy the government identified a "need to fund innovation across the public and private sectors and across the value chain from concept to market, with a key focus on high-cost development and market acceptance stages through commercialization, incubation, technology transfer and diffusion". The R&D Strategy emphasizes the need for "an approach that capitalizes on the established natural resources base while actively pursuing stronger manufacturing, information technology and biotechnology strategies", by … "connecting more effectively with the global knowledge system". Within this landscape South African institutions are grappling with new and better ways of addressing opportunities for Research and Innovation.

Rapid and continuous technological advancement of computers and telecommunications supports convergence into a single integrated system (Dickens 1998), thereby creating a competitive, volatile, dynamic, and global knowledge-focussed society. This paper reports on a project which aims to exploit this opportunity within a Triple Helix framework.

# 2 Introduction to academic intrepreneurship

## 2.1 Entrepreneurial universities

In examining the transition from research university to the entrepreneurial university Etzkowitz (2003) states that "academic enterprise is transformed in parallel, sometimes leading, other times lagging the transition to a knowledge-based economy". Etzkowitz elaborates "expectations that multi-national firms or so-called national champions will be central economic actors in the future are receding. Rather, the key economic actor is increasingly expected to be a cluster of firms emanating or at least closely associated with a university or other knowledge producing institution". Academic entrepreneurship has also expanded from an institutional growth regime into a government supported regional economic and social development strategy. Frederick Terman's initiatives to develop Stanford exemplify the stage-wise institutional evolution of universities from academic ivory towers to an innovation paradigm. By comparing five European universities: Warwick (UK); Twente (Netherlands), Strathclyde (Scotland), Chalmers (Sweden), and Joensuu (Finland) in his book 'Entrepreneurial Pathways of University Transformation', Burton Clarke (1998), highlights five requirements for transformation: "strengthened steering management core; expanded developmental periphery; diversified funding base; stimulated academic heartland; and integrated entrepreneurial culture".

## 2.2 Corporate entrepreneurs and Intrapreneurs

The 'intrapreneur' concept (Pinchot 1985) or 'corporate entrepreneur' (Burgelman 1983) is widely used in USA management literature, whereas in Europe there has been less focus on those key individuals responsible for promoting innovation and organisational change (Jones 2005). Despite intrapreneurs being frequently depicted as non-conventional individuals bent on implementing organisational change, both entrepreneurs and corporate entrepreneurs share a keen sense and ability to identify and exploit opportunities. One important distinction is that intrapreneurs usually operate as part of a team (Honig 2001).

Covin and Miles (1999) suggest two elements that define entrepreneurial organizations: 1. innovation (the introduction of a new product, process, technology, system, technique, resource or capability) and 2. ability to sustain high performance or radically improve competitiveness. Intrapreneurs must however, extend existing organisational capabilities in innovation and competitiveness without breaking links with the institution's core competencies (Floyd and Woolridge 1999). Middle managers are the locus for intrepreneurship, because they are central to resolution of the capability-rigidity paradox in corporations (Leonard-Barton 1994), through their ability to integrate knowledge and networks, by linking the operational and strategic elements of organisations activities. They create dialogue between insular departments and management, and in particular are able to identify champions responsible for operational implementation. Hitt and Ireland (2000) confirm the need for "systematic qualitative research based on ethnography, case surveys, and multi-case methods to further develop our understanding of corporate entrepreneurship in relation to strategic management of organisations". Several authors have dealt with high-technology business start-ups (Roberts 1991; Oakey 2003), but as yet the concept of Academic Intrepreneurship seems to be unfamiliar territory.

## 2.3 Academic intrepreneurship

Laukkanen, (2003) has discerned 5 types of academic entrepreneurship:

1. Large-scale science, i.e. creation of large research projects, groups or laboratories and getting them funded ('grantsmanship');
2. Supplemental income augmentation, e.g. by consulting practices, the 'lecture circuit' or private practice;
3. Industrial support for university science, by initiating, putting together and managing joint research projects or ventures;
4. Patenting, as a logical extension for commercially applicable results; and
5. Direct commercial involvement, implying formation and sole or partial ownership of firms, and, occasionally a potential use of university facilities and graduate students for the firms commercial goals, making it the most non-traditional and controversial (Louis et al. 1989).

It is this 5th category of academic entrepreneurial activity that we consider should be defined as 'Academic Intrepreneurship', since here the focus is on innovation and new venture creation.

Louis et al. (1989) report that in the first two categories individual characteristics and attitudes were the best predictors, whereas group norms are more important for commercial involvement. Jones (2005) describes the need for mobilising social capital based on open networks thereby creating opportunities to operate between rather than within network groupings. Rynes et al. (2001) have reported on the difficulties experienced in academic-management practitioner relationships in both generating and disseminating knowledge across boundaries. Structural holes created by communication gaps within open networks provide intrapreneurs opportunity to interact between departments or organisations (Burt et al. 2000). Corporate entrepreneurship and social capital concepts are brought together by the work of Chung and Gibbons (1997), who stress that while human capital concepts are understood, there is less clarity about ways in which organizational culture contributes to entrepreneurship. Hornsby et al. (2002) suggest that social capital is important to corporate entrepreneurs as it encourages risk-taking without fear of sanction. Structural hole theory is typified by competition (Burt 1992), while with social capital theory cooperation is more appropriate (Walker et al. 1997). Triple Helix models transcend these two paradigms, that of the competitive private sector (typified by confidentiality, competitive behaviour and team performance reward based systems), and the government/higher education institutional sector (characterised by public service values and individual recognition reward systems).

'Academic Intrepreneurship' takes an understanding of Academic Entrepreneurship (Laukkanen 2003) combining it with the concept of Intrepreneurship (Pinchot 1985), thereby allowing academics and students to engage in entrepreneurial and innovative endeavours within the scope of their normal activities, with support from and under the protective umbrella of a large institutional environment. In this paper we explore an Academic Intrepreneurship approach to new venture creation based on a Triple Helix model for university-industry-government research relations (Leydesdorff and Etzkowitz, 1996).

# 3 Research methods

Perry (1998) reports development of a successful, structured approach to case study methodology in postgraduate research. Sobh and Perry (2006) emphasise that realism is an appropriate epistemological guide and therefore the preferred paradigm for case study research; firstly, since it is contemporary and pre-paradigmatic requiring inductive theory building; and secondly, it does not suffer from the limitations that constructivism and critical theory do, since realism is normally characterised by some researcher objectivity. Case study research is concerned with describing real world phenomena rather than developing normative decision models, but nonetheless aims to present new theory to solve "how do" research problems.

Ogbor (2000) criticises reliance on quantitative methodologies "ostensibly based on neutral, objective and value-free social science which dominates entrepreneurship studies". Instead he advocates qualitative approaches in which there is "intimate collaboration between facts and theory". Jones (2005) argues that the researcher need not attempt to validate a theory or hypothesis but may use the case study method to illustrate ways in which corporate entrepreneurs mobilise social capital. However, it remains important to acknowledge that conceptual models are useful in mediating theory and empirical phenomena (Morgan and Morrison 1999).

Action research is appropriate for this case study as it involves institutional transformation together with the involvement of individuals affected by the change. Action research permits the dual focus of organisational development (in this instance a new university of technology), and the advancement of knowledge concerning organisational change (Hill and McGowan 1999). The case study method has been successfully used in exploring academic entrepreneurship (Brennan and McGowan 2006).

# 4 Transition strategies in Tshwane University of Technology for high-volume manufactured electronics products

## 4.1 Tshwane University of Technology

TUT was established in January 2004, through merging three former Technikons: Northern Gauteng, North-West and Pretoria. The uniquely South African designation of 'technikon' (or polytechnic) was dropped in favour of the internationally accepted 'university of technology'. TUT's Vision is 'to be the leading HEI (in Africa) with an entrepreneurial ethos …'. In support of this Vision TUT has stated as part of its Mission 'to extend the parameters of technological innovation by making knowledge useful through focused applied research and development' with a strong technological and vocational orientation.

This mega-institution annually enrols 60,000 students over an area covering four of South Africa's nine provinces. With 22 per cent of contact students accommodated in residences, TUT is by far the largest residential university in southern Africa. TUT employs over 2,700 permanent staff, including 855 highly qualified permanent academics. These academics are increasingly focusing on conducting applied research and community engagement activities in addition to their instructional roles. TUT aims to address the economic and social development of the southern African region, in its quest to promote knowledge and technology, by providing the market with a career-focused workforce.

Figure 1 shows a TUT framework for innovation and entrepreneurship, illustrated by an Innovation Fund Project entitled 'ICT networks for power optimization'. Many of the project development processes were done in the F'SATIE Incentif Incubator, and prototype and product manufacturing was done in the Technology Station in Electronics. Business development, sales and marketing of the technology, known as Net-PODSTM is left to the spin-out company Ergon Networx.

## 4.2 F'SATIE

The French South African Technical Institute in Electronics (F'SATIE) located at TUT, is a multinational DST initiative with the University of Technology of the Paris Chamber of Commerce and Industry. In response to increased demand in South Africa for engineers and lecturers in electronics, F'SATIE offers a full-time Master of Science (Electronic Engineering or Power Engineering) in collaboration with its French partner. The programme is taught by French and South African faculty as well as research and industry specialists and comprises course-work and a research project, aimed at providing participants with hardware and software skills for designing electronic systems. The research focus is on telecommunications, embedded systems, control and machine intelligence, emphasising innovation and technology management.

**Objectives**:

- Provide companies and universities with highly skilled electronics engineers and lecturers;
- Make high-quality training programs accessible to previously marginalized groups through scholarships, grants and donations;
- Holistically support students by providing them with continuous guidance, peer reviews, dedicated office space, financial support and accommodation; and
- Endorse innovation through research, industrial projects and incubation facilities.

F'SATIE is located within the Department of Electrical and Electronics Engineering of the Faculty of Engineering and the Built Environment, as depicted in Figure 1. F'SATIE offers an Incubation Centre of Technological Innovation (INCENTIF), thereby actively involving its expertise in industrial projects, and by supporting entrepreneurs in collaboration with the Technology Station in Electronics (TSE).

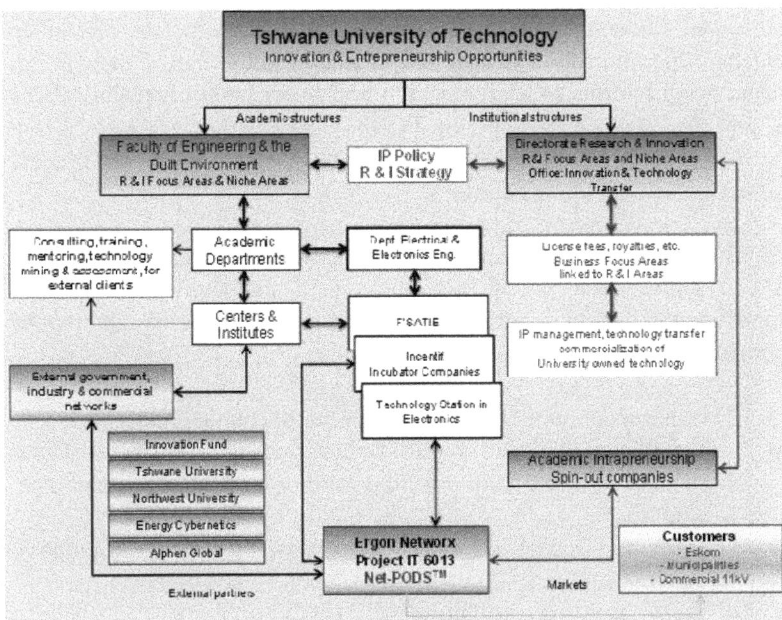


**Figure 1**: Outline of the Tshwane University of Technology environment for Academic Intrepreneurship in electronic engineering.

## 4.3 Technology Station in Electronics (TSE)

The Tshumisano Trust, meaning "cooperation or partnership" (Tshumisano, 2007), was established in 2002 by the DST as a joint venture between government, the German Agency for Technical Cooperation (GTZ) and the Committee of Technikon Principals. Its mandate is "to provide support for the Small and Medium Enterprise (SME) sector through its Technology Stations Programme". The TSP aims to strengthen activities in technological innovation and skills upgrading to increase SME competitiveness in targeted sectors including: automotive, food processing, electronics, metal value-adding, chemicals, metal casting, composite and moulded plastic. Electronics expertise available at TUT provides product development skills for SME's by engaging directly with industry. In fulfilling this mandate, the TSE was established at TUT with its primary focus to operate in the electronics, electrical and information and communications technology (ICT) industries. The TSE, in collaboration with various TUT academic departments, focuses on digital design and manufacturing of electronics prototype units, students' experiential learning, and short courses offered to industry. Operationally this requires effective and essential planning of technology platforms to align research and development activities at the TSE and the Faculty of Engineering and the Built Environment with current and future business strategies, as well as balancing product and process innovations with industry needs.

The TSE has established (with significant investment support from the DST), a sophisticated on-campus electronic manufacturing platform to support industry and academia (including student experiential training opportunities within SME's) with the following:

1 construction of prototypes and pre-production models;
2 capabilities for electronic assemblies including through-hole assembly, surface mount assembly, and re-work solutions and techniques;
3 automatic optical inspection (to improve the quality of operations);
4 short learning programmes (SLP's) for industry; and
5 experiential learning opportunities for students.

According to the Manufacturing Foresight Report (Watt 2007), manufacturing is the engine of economic growth of South Africa and must be in a healthy state to be globally competitive. The South Africa Country Report (2006) highlights that the TSP must provide skills development training to SME's to enhance their innovation capacity and competitiveness, while exposing students at universities of technology to practical challenges facing businesses. The report states that "the aims of the TSP must be to address the product and process technology needs of SME manufacturing and the service industry through technology transfer, development and diffusion and to enrich the R&D capacity of technical universities".

## 4.4 Office for Innovation & Technology Transfer (OITT)

A university in which research results are routinely scrutinised for commercial as well as scientific potential is becoming the modal academic institution (Etzkowitz 2003). Many universities now have Technology Transfer Offices (TTO's) to deal with issues relating to patenting, licensing and new venture creation based on university owned intellectual property aimed at closing the gap between invention and innovation in a Triple Helix conceptual framework. In 2004 TUT established as a part of the Directorate for Research and Innovation an Office for Innovation and Technology Transfer (OITT). The OITT has a core function to "promote an institutional culture of innovation and technology transfer". Other OITT functions are to: 1. provide consultancy and support services to staff and students, 2. implement management systems for the universities intellectual property assets, 3. contribute to the development of human potential, 4. support the establishment and activities of technology pre-incubators (on-campus) and incubators (off-campus), 5. assist with technology-based enterprise development, and 6. build international links in innovation and technology transfer.

The OITT is located in the Directorate for Research and Innovation as depicted in Figure 1. It collaborates with other TUT departments, structures and initiatives, including: Strategic Planning and Information, Legal Services, Corporate Relations and the Foundation. Over the past few years several TUT pre-incubators and Technology Stations were established including Incentif and the TSE (Engineering & Built Environment), and Technology Station in Chemicals (TSC) (Natural Sciences); the Centre for IT Product Development (CITPROD) (Information and Communications Technology); the Centre for Entrepreneurship (Management Science); the Faculty Research Committees (FRC's), and the TUT R&I Focus and R&I Niche Areas.

# 5 Outcomes

## 5.1 A spin-out company

The South African Government has established a number of innovation support programmes, most notably the THRIP Programme (Technology for Human Resources for Industry Programme) of the Department of Trade & Industry (DTI), and the Innovation Fund (IF) of the Department of Science & Technology (DST). In these programmes university staff are encouraged to partner with industry in projects aimed at creating new opportunities, and thereby contributing to South Africa's global competitiveness, job creation and regional economic development. We have previously reported on the value proposition, set up phase and management of this spin-out company (Youngleson 2006). In this paper we examine Academic Intrepreneurship processes in context of Innovation Fund Project IT 6013 entitled 'ICT for power optimisation', in which a consortium of partners including Tshwane University of Technology, Northwest University and two SME's - Energy Cybernetics and Alphen GMS formed a new business venture Pro Direct Investments 297 (Proprietary) Limited (trading as Ergon Networx) to develop, manufacture and supply to the market a novel bi-directional powerline communications (PLC) network technology know as Net-PODSTM for residential load management (RLM) and automated meter reading (AMR).

## 5.2 Academic outcomes

TUT academic outcomes can be classified according to the Workload norms/guidelines (v8.027) approved by Senate on 05/03/2007. TUT academics work on the following activities: Teaching and Learning, Research and Innovation (R&I), Community Engagement, and Administration. As this paper falls largely within the R&I work category, the other categories are not discussed.

R&I work category outputs are defined as:

1 Direct outputs - published articles, proceedings, books, artefacts, artistic exhibitions, and patents;
2 Conference attendance - (where an paper is presented), conferences held at TUT; local or international;
3 R&I Focus and Niche area involvement;
4 Consultation - as a researcher, and/or supervision and assisting postgraduate students (Masters & Doctorate) in preparation of final article(s), and theses;
5 External funding - research grants, industry collaboration, research fellowships.

Ten staff members were directly involved in Project IT 6013, eight from Tshwane University of Technology, two from Northwest University.

**Table 1:** Summary of outputs in Project IT 6013 in the areas designated

| **Teaching & Learning** | **Academic outcomes** | **Quantity** |
|---|---|---|
| | University students trained (Diploma & Degrees) | 19 |
| | ESCO personnel trained (SLP) | 12 |
| **Research & Innovation** | Research & Innovation outcomes | |
| | Products & Prototypes (hardware, software, test system) | 29 |
| | Patents (and provisional patents), registered designs, trademarks | 8 |
| | International conferences attended (papers presented) | 1 |
| | Local conferences attended (papers presented) | 3 |
| | Participation in TUT R&I Niche Areas (Management of Innovation, and Business Development) | 2 |
| **Community Engagement** | Community engagement | |
| | Pilot site projects | 5 |
| | ESCO training projects | 3 |
| | Schools expo on energy & electricity with Tshwane Municipality | 1 |
| **Administration** | Reports and plans | |
| | Detailed Project Plan | 1 |
| | Quarterly Reports | 12 |
| | Mid-term progress reports | 1 |
| | Interim commercialisation plan | 1 |
| | Final Report | 1 |
| | Commercialisation plan | 1 |
| | Business plan | 1 |
| | Total Project IT 6013 outputs | 101 |

## 5.3 Financial summary

**Table 2:** Innovation funds received by TUT for Project IT 6013

| **Year 1** | **Year 2** | **Year 3** |
|---|---|---|
| R 4,404,384 | R 4,438,006 | R 3,790,881 |
| R 6 = US$ 1 | TOTAL: | R 12,633,271 |

## 5.4 Intellectual property

**Table 3:** Patenting and registration of designs in project IT 6013

| Patent / Design | Country | Number | Filing Date | Title | Status |
|---|---|---|---|---|---|
| Patent | PCT | PCT/ZA2006/008 | Jan-06 | Temperature Control: Net-PODS | Due Date 23 February 2008 |
| Design | South Africa | F2007/00171 | Feb-07 | Enclosures | Pending |
| Design | South Africa | F2007/00172 | Feb-07 | Lids | Pending |
| Design | South Africa | F2007/00173 | Feb-07 | Lids | Pending |
| Provisional Patent | South Africa | 2007/01596 | Feb-07 | Data Communications | Due Date 23 February 2008 |
| Provisional Patent | South Africa | 2007/01597 | Feb-07 | Temperature Measurement | Due Date 23 February 2008 |

## 5.5 Other outcomes

Project IT 6013 exposed a number of students at various academic levels at TUT to a commercialisation environment from concept to final product. TSE staff experienced an opportunity to gain valuable experience and to mentor students and other staff members on manufacturing optimisation, prototype construction, machine programming and usage, to prepare manufacturing documentation, machine and hand assembly techniques, and testing of the manufactured products.

The product developed in Project IT 6013, Net-PODSTM, further contributes towards experiential training at the TSE as well as providing valuable inputs for curriculum refocusing of the experiential training component for future students. Most importantly, this project laid the foundation for similar commercialisation efforts within TUT at the Faculty of Engineering and the Built Environment. Recently, Ergon Networx has commenced negotiations with large-scale manufacturers of electronics products, thereby creating capacity and knowledge required to take the product Net-PODSTM to the next level of manufacturing.

# 6 Problems and challenges

## 6.1 Manufacturing considerations

Critical resources supporting commercialisation include: good infrastructures, support systems, and applicable technology platforms operated by skilled staff thereby directly influencing the design, development and commercialisation process.

The success of this project was made possible by the team effort and collaboration between academic departments and non-academic departments within TUT such as the TSE, the Incentif Incubator, the two SME's (Energy Cybernetics and Alphen GMS) and the two universities involved (TUT and Northwest) in the design, development and commercialisation process. This project highlights the challenge of availability and quality of the staff (employees) and the suitability of support and development services that can foster product commercialisation. Limited availability of these factors might prejudice it or even inhibit progress.

Challenges for TUT from an electronics product commercialisation perspective:

1. reliable and available support services;
2. availability of development, manufacturing and test services;
3. content of the support and development services;
4. availability and maintenance of well equipped infrastructure; and
5. availability of skilled staff to perform the commercialisation process.

## 6.2 Company structures and administration

This project was very unusual from a management and functioning point of view. Ergon Networx did not exist as a going concern before the project started, and thus represents a true 'spin-out' company. This company was supported and staffed by members of the initial consortium. To run a successful multidisciplinary project needs all the supporting structures that would normally exist within an organization. When the project began early in 2004 the only framework that existed were the Innovation Fund grant agreement and company shareholder agreement, and the Project IT 6013 plan. The company was formed right at the outset, to manage the project, as a legal framework for cooperation between the consortium members, to own all the intellectual property, and to commercialise the technology at the appropriate time.

The systems used for running the project and business were sometimes confusing and difficult to work with:

1. The consortium was made up of several different parties with different needs;
2. Strict control on disbursement of funds was required in terms of the grant agreement. A paper trail was required to ensure that only people with the necessary authority signed the approval documentation for any payment. Initially a cheque system with at least two signatories – either Managing Director / Project Director / or Chairman was used (later electronic payments were introduced);
3. The Innovation Fund paid the project funds directly to TUT, who then paid Ergon Networx against invoice for completed deliverables; Ergon Networx paid subcontractors, or suppliers. The process from the time of approval of a quarterly IF report to the disbursement of funds could take three or more months.

University bureaucracy often slowed internally generated project intensity (the hurry up and wait syndrome). A solution is not easily found to these complicated arrangements. Multidisciplinary projects are by nature complex from a management perspective, and require time and effort to keep on track. It is also challenging to do business development simultaneously as part of a technical development project, but this example is unlikely to be completely unique.

## 6.3 Main lessons learned

It is impossible to report on all the lessons learned by everyone involved in this complex project. We report on the main lessons learned below:

1. Technological problems may arise without warning;
2. Scope creep is a constant challenge, severely impacting budgets and delaying next phases;
3. Potential customers determine the pace at which they are willing to move forward;

4 Business development is a fluid process, it is difficult to be dynamic within a defined project management structure;
5 Strategy must constantly adapt to changing market conditions;
6 The project plan provided structure to the effort, and desired project outcomes were clearly defined, and attained;
7 Provision was made for company structures and personnel, and for capital infrastructure investment;

# 7 Conclusions

Universities are not in the business of starting businesses, neither is it their core mission to manage start-up companies. They are however, part of the larger community, and their third mission encompasses the need to contribute to the regional economic development. The strategic decision by TUT to become a university of technology with an entrepreneurial ethos, requires a different set of norms, infrastructure, culture and behaviour to align itself with regional and/or industry needs. New structures are required, such as an Office for Innovation and Technology Transfer and on-campus incubators and technology stations. Joint projects, project and financial management skills, alternative funding models, intellectual property and licensing issues become important considerations.

Some academics who are used to a traditional university culture, may find this new environment contradictory, undisciplined and threatening. There are however, many who see the transition to a more entrepreneurial university culture as inevitable, to be faced and controlled, and contributing to the university research and innovation effort. Typically, there is a lack of infrastructure, skills, and resources required to support Academic Intrepreneurship, although strategies to address this shortage are emerging to describe university spin-outs such as 'skunk-works', 'virtual entrepreneurship', and 'temporary entrepreneurial positions' (van Alsté and van der Sijde 1998).

Most team-based approaches to intrapreneurial innovation focus on improving processes, services and products yet remain at the lowest level, that of incremental improvement. The case presented here demonstrates an institutional approach that addresses the higher-order goals of high-technology product innovation. A correct and supporting environment with appropriate knowledge development at the individual, team and institutional level provides a true action learning approach capable of providing innovations through Academic Intrepreneurship.

The authors have presented the micro-level case of an on-campus development and manufacturing case within the electronics product development sector. It relates directly to the improvement of product development strategies in a university setting, thereby contributing to the success of the institution from a number of perspectives – education of students, technological and social sciences research outcomes, generation of new intellectual property, and formation of an income generating spin-out Company in which the university will enjoy profit sharing. At the macro level this experience contributes to the ability to inform policy formulation and strategy and to improve management decision-making.

## Acknowledgements

The authors wish to thank and acknowledge the Innovation Fund Trust, of the DST; Technology Stations Programme (Tshumisano), of the DST; Tshwane University of Technology; Pro Direct Investments 297 (Pty) Ltd., trading as Ergon Networx.

## References

Ballard, S., James, T., Adams, T., Devine, M., Laysa, L. and Meo, M. (1989) Innovation through technical and scientific information: Government and industry cooperation. New York: Quorom.

Brennan, M.C. and McGowan P. (2006) Academic entrepreneurship: an exploratory case study, International Journal of Entrepreneurial Behaviour & Research, Vol 12, No. 3, pp 144-164.

Burgelman, R.A. (1983) Corporate entrepreneurship and strategic management: insights from a process study, Management Science, Vol 29, pp 1394-64.

Burt, R. (1992) Structural Holes: The Social Structure of Competition, Harvard University Press, Cambridge, MA.

Burt, R., Hogarth, R. and Michaud, C. (2000) The social capital of French and American managers, Organisation Science, Vol 11, No. 2, pp 123-147.

Chung, L.H. and Gibbons, P.T. (1997) Corporate entrepreneurship: the roles of ideology and social capital, Group and Organization Management, Vol 22, No. 1, pp 10-30.

Clark, B.R. (1998) Creating Entrepreneurial Universities – Organizational Pathways of Transformation, Issues in Higher Education, Pergamon, Oxford.

Covin, J. and Miles, M. (1999) Corporate entrepreneurship and the pursuit of competitive advantage, Entrepreneurship: Theory and Practice, Vol 23, No. 3, p 47.

DACST – The Department of Arts, Culture, Science and Technology (2002), South Africa's National Research and Development Strategy, available at http://www.info.gov.za/otherdocs/2002/rd_strat.pdf

Dickens, P. (1998) Global Shift – Transforming the World Economy, 3rd ed., PCP, London.

Etzkowitz, H. (2003) Research groups as 'quasi-firms': the invention of the entrepreneurial university, Research Policy, Vol 32, pp 109-121.

Floyd, S. and Woolridge, B. (1999) Knowledge creation and social networks in corporate entrepreneurship: the renewal of organisational capabilities, Entrepreneurship: Theory and Practice, Vol 23, No. 3, p 123-145.

Hage, J. and Hollingsworth, J.R. (1997) Idea-innovation networks: Integration Institutional and organizational levels of analysis. EMOT conference, Berlin, 30 January 1997.

Hill, J. and McGowan, P. (1999) Small business and enterprise development: questions about research methodology, International Journal of Entrepreneurial Behaviour and Research, Vol 5, No. 1, pp 5-18.

Hitt, M. and Ireland, D. (2000) The intersection of entrepreneurship and strategic management, in Sexton, D. and Landstrom, H. (Eds), Handbook of Entrepreneurship, Blackwell, Oxford.

Honig, B. (2001) Learning strategies and resources for entrepreneurs and intrapreneurs, Entrepreneurship: Theory and Practice, Vol 25, No. 3, pp 21-35.

Hornsby, J., Kuratko, D. and Zahra, S. (2002) Middle managers' perception of the internal environment for corporate entrepreneurship: assessing a measurement scale, Journal of Business venturing, Vol 17, pp 253-273.

http://www.un.org/esa/agenda21/natlinfo/countr/safrica/atmosphere.pdf

Jones, O. (2005) Manufacturing regeneration through corporate entrepreneurship: middle managers and organizational innovation, International Journal of Operations & Production Management, Vol 25, No. 5, pp 491-511.

Laukkanen, M. (2003) Exploring academic entrepreneurship: drivers and tensions of university-based business, Journal of Small Business and Enterprise Development, Vol 10, No. 4, pp 372-382.

Leonard-Barton, D. (1994) Core capabilities and core rigidities: a paradox in managing new product development, Strategic Management Journal, Vol 13, pp 111-125.

Leydesdorff, L. and Etzkowitz, H. (1996) Emergence of a Triple Helix of university-industry-government relations, Science and Public Policy, Vol 23, pp 279-86.

Louis, K.S., Blumenthal, D., Gluck, M.E. and Stoto, M.A. (1989) Entrepreneurs in academe: an exploration of life scientists, Administrative Science Quarterly, Vol 34, No. 1, pp 110-131.

Morgan, M.S. and Morrison, M. (Eds) (1999) Models as Mediators: Perspectives on Natural and Social Sciences, Cambridge University Press, Cambridge.

Oakey, R. (2003) Technical entrepreneurs in high technology small firms: some observations on the implications for management, Technovation, Vol 23, No. 8, pp 679-691.

Ogbor, J.O. (2000) Mythicizing and reification in entrepreneurial discourse: ideology-critique of entrepreneurial studies, Journal of Management Studies, Vol 37, No. 5, pp 605-635.

Perry, C. (1998) Processes of a case study methodology for postgraduate research in marketing, European Journal of Marketing, Vol 32, No. 9/10, pp 785-802.

Pinchot, G. III (1985) Intrapreneuring: Why you don't have to leave the Corporation to become an Entrepreneur, Harper and Row, New York, NY.

Roberts, E.B. (1991) Entrepreneurs in High-Technology: Lessons from MIT and Beyond, OUP, Oxford.

Rynes, S.L., Bartunek, J.M. and Daft, R.L. (2001) Across the great divide: Knowledge creation and the transfer between practitioners and academics. Academy of Management Journal, Vol 44, No. 2, pp 340-355.

Sobh, R. and Perry, C. (2006) Research design and data analysis in realism research, European Journal of Marketing, Vol 40, No. 11/12, pp 1194-1209.

South African Country Report, (2006), available at: Tshumisano. (2007), available at: http://www.tshumisano.co.za.

van Alsté, J.A and van der Sijde, P.C. (1998) The role of the university in regional development: the University of Twente. Enschede. Twente University Press.

Walker, G., Kogut, B. and Shan, W. (1997) Social capital, structural holes and the formation of an industry network, Organisational Science, Vol 8, pp 109-125.

Watt, P. (2007) Foresight Manufacturing Report, Department of Science & Technology (DST), available at: http://www.dst.gov.za/

Youngleson, J.S. (2006), Unlocking the innovative potential embedded in higher education institutions: a case study of a multidisciplinary collaborative triple helix approach, presented at the 29th ISBE Conference, Cardiff, Wales.

# A Method for Monetizing Technology Innovations

**Arcot Desai Narasimhalu**
Singapore Management University, Singapore

**Editorial commentary**

Narasimhalu discusses the contributions universities can make to transforming technological innovations that have emerged from academic research projects into commercially viable products, and assesses the challenges and dilemmas they typically face when attempting to embark on commercialisation ventures. The author proposes a method for monetising technological innovations by means of the careful assessment of such elements as innovation properties and value drivers. Underlying the case is a discussion of a common tension between the matching of the properties of technological innovations with those of business innovations.

The method was developed through extensive study and observation of commercialisation processes.

Some points for discussion and learning arising from this case include:

- The role of universities in the commercialisation of technological innovations.
- The processes involved in transforming a technological innovation into a commercially viable product.
- Approaches for successfully matching innovations with markets.
- The tensions between academic freedom and entrepreneurial necessity.

*Arcot Desai Narasimhalu*

**Abstract**: Technology innovations from most universities and research institutes are generally created with no clear path to commercialization in mind. This is largely due to the culture in academic institutions and research institutes whose mission it is to explore the creation of innovations that promise long term benefits. This culture of academic freedom leads to a stockpile of technology innovations at their technology transfer offices (TTOs). These offices are often in a dilemma on how best to monetize the technology innovations that are in their custody. While there have been many social science research methodology based studies on this subject under the broad umbrella of "Technology Management", there is still no method available to help TTOs manage the commercialization of their IP better. A clearly articulated method for translation of technology innovations into business innovations will certainly help move accumulated IP at the TTOs to the markets. Such a method should help identify the best commercial application of technology innovations. This paper is based on action or practice research. This approach uses empirical data and experiential observations derived from several years of commercialization experience to derive and define a method. The method will present a framework for capturing the properties of technology innovations. A heuristic algorithm is used to achieve a prioritized list of business innovations that can be generated from a set of technology innovations. The paper will use an example to illustrate the method. An early version of the algorithm is being used in a course on technopreneurship. This exercise is expected to both validate and refine the method presented in the paper.

**Keywords**: Technology innovation, business innovation, technology transfer office, commercialization, optimal

# 1 Introduction

From time to time one observes that a technology innovation that was created for one market gets adopted in a different and unexpected market on a wider scale. An example is the steam engine. Steam engine was first envisioned for stationary use - pumping water from a coal mine to the surface by Thomas Savery in 1698 based on Denis Papin's Digester or pressure cooker of 1679 (innovators.about.com). It however had the maximum impact in mobile use – various forms of locomotives.

The above example shows that a technology innovation can be used to create more than one business innovation. A challenge often faced by the inventor or the organization managing technology innovations is to figure out the best strategy for monetizing their suite of technology innovations. There are several proprietary practices and these are often ad-hoc and not disclosed to the world at large as best or preferred practices. This paper will describe an approach to understanding the properties of value drivers that customers desire, the properties exhibited by technology innovations and use these set of properties to derive a method for selecting the best means of monetizing technology innovations.

Section 2 of this paper presents a literature review. Section 3 presents the proposed method and Section 4 contains discussions and conclusions.

## 2 Literature review

There have been a number of studies related to innovation (Abernathy and Utterback (1978), Kelly and Kranzber (1978), Leifer, Colarelli and Peters (2000), Schumpeter (1934, 1939, 1942), Trott (1998), Utterback (1994), Van de Ven (1999) and von Hippel (1978)). These papers have looked at different aspects of innovation. They have been followed by innovation models as described in Rothwell (1992) and presented in Table 1. In the last decade there have been several methodologies to identify innovation opportunities (Christensen (1997), Kim and Mauborgne (2005), La Salle (2002, 2006) and Narasimhalu (2005) and some are listed in Table 2.

Table 1: Different innovation models

| Year | Innovation Model |
|---|---|
| 1950s / 60s | Technology Push Model |
| 1970s | Market Pull Model |
| 1980s | Coupling Model |
| 1980s/ 90s | Interactive Model |
| 2000s | Network Model |

Table 2: Recent work in innovation identification methodologies

| Year | Innovation Methodology |
|---|---|
| 1997 | Disruptive Innovation |
| 1997 | Value Innovation |
| 2002 | Thinking Matrix |
| 2005 | Innovation Cube |
| 2006 | La Salle Innovation Matrix |

All of the above studies have contributed significantly towards advancing the understanding, identification and management of innovation opportunities. However, none of them has critically examined the properties that technology innovations share with value drivers considered to be important to customers. This paper fills this void by presenting a property based approach towards finding the best means of realizing value from a set of technology innovations.

# 3 The proposed method

This section first presents the definitions used in the proposed method. The proposed method matches the properties of technology innovations with the properties of business innovations to arrive at a suggested means of monetizing technology innovations. A sample list of properties of technology innovations is presented in section 3.2. The value drivers of innovations and their properties are discussed in section 3.3. Section 3.4 introduces TechVal matrix that is used to identify the path to monetizing technology innovations in section 3.5. This section also discusses the proposed method using an example.

## 3.1 Definitions

This section provides definitions for terms used in the paper.

*Addressable market* – is defined to be the product of the number of estimated customers times the price they are willing to pay for the business innovation. This number is an estimate.

*Business Innovation* – is defined to be a product, solution, application, platform, service or any other process which meets either the explicit or the implicit demands of the customers in a given market. Business innovations are often built using technology innovations. Some business innovations use technology innovations. Other business innovations are built on existing or mature technologies to address a new market opportunity.

*Innovation Drivers* - are either the pains or the demand for enhanced experience (also called pleasure) by a group of customers. Innovation drivers define the opportunities for creating successful business innovations. An Innovation Driver may lead to one or more Business Innovations. An example can be body ache. This is a pain and hence an innovation driver. This can result in a foot massaging equipment, a business innovation. It could also result in a separate neck massaging equipment, another business innovation. There could also be a back massager, a third business innovation. All these three may get integrated into a total massaging solution, a fourth business innovation.

*Innovation Triggers* – are defined to be market shifts and / or a technology shifts that creates opportunities for successful business innovations. Regulations or deregulations are good examples of market shifts. Emergence of new group of users for a product is another example of a market shift. Market shifts can be represented as Innovation Rules or Innovation Chains (Narasimhalu (2005)). Multi-touch technology is an example of a technology shift. A technology shift can result in the creation of one or more successful business innovations. Universal Serial Bus or USB is a great example of a technology shift that resulted in several business innovations including Thumb drives TM.

*Innovation Rule / Chain* - is a representation of observed phenomenon on how a market or a business innovation evolves over time. Computers were first created for Defence applications and later lead to the creation of main frames, mini computers, personal computers, lap tops and now personal digital assistants or PDAs. This evolution can be captured as an Innovation Chain (Narasimhalu 2005)). An innovation chain with a single link is called an innovation rule. Each link in an Innovation Chain defines a new market opportunity. A business innovation that currently addresses an enterprise market is most likely to shift to a consumer market or vice versa when the required technology shift takes place.

*Technology Innovation* – is defined to be a new technology that provides an improvement over one or more existing technologies. Examples of improvements could be either reduced cost or increased functionality. Technology innovations will be a means for creating new value for customers.

*Value Drivers* – are defined to be the properties of business innovations that address innovation drivers. For example, if the pain of standing in long queues at a cashier's register is an innovation driver, faster transaction processing is the corresponding value driver. Either a larger number of service stations or a new solution to speed up the ringing in the purchases and completing the payment will be the corresponding business innovations. The first alternative will be a service innovation and the second alternative will be a product innovation. Service innovations may not always require innovations in technology.

## 3.2 Properties of technology innovations

The following are a sample set of categories of properties of technology innovations. A technology innovation could have one or more properties from across different categories. Every organization should continually maintain a list of properties for its technology innovations. These properties will be used in the method to monetize technology innovations.

***Functional properties***

- TF1: Cost reduction
- TF2: Increased productivity
- TF3: Faster processing
- TF4: Increased display quality
- TF5: Reduced size
- TF6: Increased power

***Emotional properties***

- TE1: Improved look and feel
- TE2: Ease of use
- TE3: Substitute for animal testing
- TE4: Environment friendliness
- TE5: Choice of colours

***Communal properties***

- TC1: Personal improvement
- TC2: Interpersonal communications
- TC3: Intercommunity communications
- TC4: Intra-enterprise communications
- TC5: Groupware

These three categories are examples. There can certainly be more than these three categories. The examples given under each of the categories are by no means exhaustive. A firm or an organization should develop a customized list of categories and subcategories of properties of technology innovations in its custody.

The first step is to derive a mapping between technologies and their properties.

Each property of a technology innovation is labelled TXn where T stands for technology innovation; X is a unique letter indicating the category the property belongs to and n is a unique number assigned to the property. The sample properties listed above get labelled TF1 through TF6, TE1 through TE5 and TC1 through TC5.

Table 3 is an example using the property categories of technology innovations defined above. T1 through Tn are the technologies under consideration. A technology innovation can have one or more properties. The contents of the table are for the purposes of illustration only and do not represent real technologies.

**Table 3:** Mapping between technology innovations and their properties

| | Technologies | | | | | | | | | | | | |
|---|---|---|---|---|---|---|---|---|---|---|---|---|---|
| **Properties** | T1 | T2 | T3 | T4 | | | | | | | | Tn-1 | Tn |
| | TF1 | TE1 | TF3 | TC1 | | | | | | | | TC1 | TC1 |
| | TE3 | TE2 | TF4 | TF4 | | | | | | | | TC2 | TE2 |
| | TC4 | TE3 | TF5 | TE3 | | | | | | | | TC3 | TF5 |
| | TC5 | TE4 | TF6 | | | | | | | | | TC4 | TF6 |

## 3.3 Value drivers or properties of business innovations

Recall value drivers are the properties of business innovations that customers will be willing to pay for. Every organization should develop its value driver categories and the value drivers under each of the categories to represent the emerging market demands. These market demands and could be generated using Innovation Rules or Innovation Chains or any other means. Each of the value drivers should be labelled VXn where V stands for value created; X is a unique letter representing the Value Driver category and n is a unique number representing a property within each subcategory. Table 4 shows an example of the link between innovation drivers and corresponding value drivers.

Table 4: Value drivers or properties demanded by innovation drivers

| Value Driver Category | Innovation Driver Category | Value Driver 1 | Value Driver 2 |
|---|---|---|---|
| Functional | Long waiting times | Faster processing times (VF1) | Increased service capacity (VF2) |
| | Not affordable | Cheaper product / service (VF3) | Low cost components (VF4) |
| | Too bulky to carry | Miniaturization (VF5) | Consolidation of components into one (VF6) |
| | Need the product / service anytime anywhere | Mobile connectivity (VF7) | Miniaturization (VF8) |
| | Difficult to use | Better user interface (VF9) | Simplified interface (VF10) |
| | Too complex | Reduced feature set (VF11) | |
| Emotional | Better quality TV viewing experience | Larger screen size (VE1) | Higher resolution (VE2) |
| | More comfortable seating in cinemas | Bigger seats (VE3) | Reclining seats (VE4) |
| | Better product design | Improved look and feel (VE5) | |
| | Better entertainment | Improved audiovisual experience (VE6) | Improved programming (VE7) |
| | Be more presentable | Burn fat (VE8) | Accessories (VE9) |
| Communal | Simultaneous audio visual communication with several people | High quality multiple video stream transmission (VC1) | Managing multiple sessions within a window (VC2) |
| | Easy group discussion | More than one display for a laptop (VC3) | |
| | Multiplayer entertainment | Middleware to support multiple players (VC4) | Middleware to support multiple games (VC5) |

It is important to note that the categories for properties of technology innovations can be different from the categories for value drivers. Value drivers for two different innovation drivers may be same, for example VF5 and VF8. Not all value drivers will require a technology based solution. The value driver VE3 bigger seats, does not require any new technology. Again, the value driver VF2 requires increased capacity provisioning and not new technologies. Some of the value drivers will of course require the use of technology innovations with matching properties.

## 3.4 TechVal matrix

There are two key principles used in developing this method.

Key principle 1: The properties demanded by a value driver may be fulfilled by the properties of one or more technology innovations that collectively satisfy the requirements at the lowest cost.

Key principle 2: A technology innovation can satisfy more than one property demanded by one or more value drivers.

Let the cost of technology innovation Ti be CTi.

Let the Addressable market for a Business Innovation be A(Bi). A business innovation might be made up of several value drivers V1, V2, ..., Vk.

Let Rj, be estimated revenue from the sale of one unit of 'j'th business innovation. Rj can be defined as:

Rj = Sum (RXi), where RXi is the estimated revenue for providing value driver VXi.

The revenue for a value driver will be proportional to the demand for it. If a demand for a value driver is high then the revenue or the money a customer is willing to pay will be correspondingly high.

A(Bi) can be calculated using a formula such as the one shown below. If Ni is the estimated unit sales of 'i'th business innovation, then the addressable market of the 'i'th business innovation can be derived as,

A(Bi) = Ni x Ri

TechVal Matrix represents Innovation Drivers (IDi), their corresponding Value Drivers (VXn), Technology innovations (Tj) and their properties (TXm) as shown in Table 5.

**Table 5:** TechVal matrix for the properties discussed in sections 3 and 4

| Innovation Drivers | Value Drivers | RXi | N | A (Bi) | Technology innovations and their Properties | | | | | | | |
|---|---|---|---|---|---|---|---|---|---|---|---|---|
| | | | | | Tech 1 | | Tech 2 | | | Tech 3 | | Tech 4 |
| | | | | | CT1 | | CT2 | | | CT3 | | CT4 |
| | | | | | TF 1 | TF 2 | TC 1 | TC 2 | TC 3 | TF 5 | TF 6 | TE3 |
| ID1 | VF1 | RF1 | N 1 | A1 | X | | | | | | | |
| | VE1 | RE1 | | | | | | | | | | X |
| ID2 | VF3 | RF3 | N 2 | A2 | | | | | | | | |
| ID3 | VC2 | RC 2 | N 3 | A3 | | | | X | | | | |
| | VF5 | RF5 | | | | | | | | | X | |
| | VE3 | RE3 | | | | | | | | | | |
| ID4 | VF7 | RF7 | N 4 | A4 | | | | | | | X | |

*An 'X' placed in any of the cells of a TechVal matrix indicates a match between the property of a technology innovation and a value driver.*

## 3.5 Method for monetizing technology innovations

TechVal matrix certainly helps match properties of technology innovations with the properties with value drivers and forms the building block of the proposed method.

### *3.5.1 The method*

1. Identify the technology innovations under consideration.
2. Define the categories for representing the properties of technology innovations.
3. List the properties of each of the technology innovations.
4. Generate a list of known Innovation Drivers. For the sake of simplicity assume each innovation driver leads to one business innovation.
5. Forecast the sales estimates of quantities of each of the Innovation Drivers.
6. Derive a list of Value Drivers for each of the Innovation Drivers.
7. Estimate the monetary value of each of the Value Drivers.
8. Project the sales forecast for each of the Business Innovations.

9 Determine the addressable market size for each of the value drivers (optional if needed).
10 Calculate the addressable markets of each Business Innovation / Innovation Driver.
11 Construct TechVal matrix for the set of chosen Innovation Drivers and Technology innovations.
12 Select the Value Driver with the (next) largest addressable market. Identify sets of technology innovations whose properties match the property demanded by the Value Driver.
13 Pick the set of technology innovations which cost the least while still satisfying the needs of Innovation drivers.
14 Iterate steps 12 and 13 until all Value Drivers have been considered.

The technology innovations whose properties match the properties of the value drivers and are the least cost would be selected through this method. If a technology innovation is picked only for one value driver and the addressable market for that value driver is very large then it may be best to build a company around this technology innovation. If a technology innovation is picked against multiple value drivers, it is likely to be a natural candidate for licensing.

### 3.5.2 An example

The following steps illustrated the use of the method with the aid of an example.

1. Identify the technology innovations under consideration.

- T1 – MP3 Coder / Decoder
- T2 – USB Disks
- T3 – High resolution LCD displays
- T4 – Light weight security protocols

2. Define the categories for representing the properties of technology innovations.

***Communal***

- TC1 – Easy sharing of content
- TC2 – Portable content

***Emotional***

- TE1 – Sharper picture viewing
- TE2 – Privacy and Security

***Functional***

- TF1 – Digitized music
- TF2 – New standard adopted by many
- TF3 – Increased storage capacity
- TF4 – Better viewing in smaller formats

3. List the properties of each of the technology innovations.

- T1 – TF1, TF2
- T2 – TC1, TC2, TF3
- T3 – TE1, TF4
- T4 – TE2

4. Generate a list of known Innovation Drivers. For the sake of simplicity assume each innovation driver leads to one business innovation.

- ID1 – Cannot listen to music when I want where I want
- ID2 – No displays on portable devices
- ID3 – Limited functions on mobile phones
- ID4 – Do not want others to access my portable data

5. Forecast the sales estimates of quantities of each of the Innovation Drivers.

- ID1 – 10; ID2 – 50; ID3 – 40; ID4 - 50

6. Derive a list of Value Drivers for each of the Innovation Drivers.

***Value Drivers for ID1***

- VF1 – Portable music player
- VE1 – Prefer graphics matching the music
- Value Drivers for ID2
- VF2 – Displays on small devices

***Value Drivers for ID3***

- VC1 – High resolution screens to support good quality video conference
- VF3 – Removable storage
- VE2 – Listen to different music collections
- Value Drivers for ID4
- VF4 – Secure my portable data

7. Estimate the monetary value of each of the Value Drivers.

RXi – The price customers are willing to pay for the value driver. The numbers are in dollars and for illustrative purposes only. Monetary values of the different value drivers are listed below.

***Functional value drivers***

- VF1 – 10 dollars
- VF2 – 10 dollars
- VF3 – 20 dollars
- VF4 – 50 dollars

***Emotional value drivers***

- VE1 – 10 dollars
- VE2 – 10 dollars

***Communal value drivers***

- VC1 – 30 dollars

8. Forecast the sales for each of the business innovations.

Ni – Estimated number of units of a business innovation sold in millions.

Estimated numbers of 10 million MP3 players and 40 million mobile phones are for illustrative purposes only.

- BI1 corresponds to ID1and is meant for MP3 players only – 10 million
- BI2 corresponds to ID2 and is meant for mobile phones and MP3 players – 50 million
- BI3 corresponds to ID3 and is meant for mobile phones only – 40 million
- BI4 corresponds to ID4 and is meant for mobile phones and MP3 players – 50 million

9. Calculate the addressable markets of each Business Innovation / Innovation Driver.

Business Innovation Bi is in response to Innovation Driver IDi.

A(Bi) – Addressable market in millions of dollars. This value is derived from estimated numbers and the price the customers are willing to pay for a given value driver. The value for each Innovation Driver is determined by the sum of the products of the price customers are expected to pay for a value driver times the market size in terms of the number of units sold. The sum is only for the value drivers satisfied by the suite of technologies. If a given suite of technologies do not meet the needs of a value driver then the product of its value times the estimated market size should not be included in the calculations.

- A(B1) = (RF1 + RE1) N1 = (10+10) x 10 = 200 million dollars.
- A(B2) = RF2 x N2 = 10 x 50 = 500 million dollars.
- A(B3) = (RC1 + RF3 + RE2) x N3 = (30+20+10) x 40 = 2400 million dollars.
- A(B4) = RF4 x N4 = 50 x 50 = 2500 million dollars

10. Construct TechVal matrix for the set of chosen Innovation Drivers and Technology innovations. See Table 6.

**Table 6:** Techval matrix for the example

| Inno-vation Drivers | Value Drivers | RXi $ | Ni | A(Bi) | Technology innovations and their Properties | | | | | | | |
|---|---|---|---|---|---|---|---|---|---|---|---|---|
| | | | | | T1 MP3 coder/ decoder | | T2 USB disks | | | T3 High resolution LCD displays | | T4 Light weight security protocols |
| | | | | | CT1=10 | | CT2=10 | | | CT3=20 | | CT4=30 |
| | | | | | TF1 | TF2 | TC1 | TC2 | TF3 | TE1 | TF4 | TE2 |
| ID1 | VF1 | 10 | 10 | 200 | | | | | | | | |
| | VE1 | 10 | | | | | | | | | | |
| ID2 | VF2 | 10 | 50 | 500 | | | | | | | | |
| ID3 | VC1 | 30 | 40 | 2400 | | | | | | | | |
| | VF3 | 20 | | | | | | | | | | |
| | VE2 | 10 | | | | | | | | | | |
| ID4 | VF4 | 50 | 50 | 2500 | | | | | | | | |

11. Select the Value Driver with the (next) largest addressable market. Identify sets of technology innovations whose properties match the property demanded by the Value Driver. See Table 7.

**Table 7:** TechVal matrix for the examples with matches identified

| Inno-vation Drivers | Value Drivers | RXi $ | Ni | A(Bi) | Technology innovations and their Properties | | | | | | | |
|---|---|---|---|---|---|---|---|---|---|---|---|---|
| | | | | | T1 MP3 coder/ decoder | | T2 USB disks | | | T3 High resolution LCD displays | | T4 Light weight security protocols |
| | | | | | CT1=10 | | CT2=10 | | | CT3=20 | | CT4=30 |
| | | | | | TF1 | TF2 | TC1 | TC2 | TF3 | TE1 | TF4 | TE2 |
| ID1 | VF1 | 10 | 10 | 200 | X | X | X | X | X | | | |
| | VE1 | 10 | | | | | | | | X | | |
| ID2 | VF2 | 10 | 50 | 500 | | | | | | | X | |
| ID3 | VC1 | 30 | 40 | 2400 | | | | | | X | X | |
| | VF3 | 20 | | | | | | X | | | | |
| | VE2 | 10 | | | | | | X | X | | | |
| ID4 | VF4 | 50 | 50 | 2500 | | | | X | | | | X |

- ID4 has the largest addressable market.
- ID3 has the next largest addressable market.
- ID2 has the next largest addressable market.
- ID1 has the next largest addressable market.

12. Pick the set of technology innovations which cost the least while still satisfying the needs of the innovation drivers.

CTi – cost of technologies. They will come into play when more than one technology satisfies the value drivers.

- Only one set of technologies T2 and T4 has a property match. Select them.
- Only one set of technologies T2 and T3 has a property match. Select them.
- Only one set of technology T3 has a property match. Select it.
- Only one set of technologies T1 and T2 has a property match. Select them.

13. Iterate steps 11 and 12 until all Value Drivers have been considered.

Notes:

- T4 is used in only one innovation driver and has the largest revenue potential and is hence perhaps a serious candidate for spin-off.
- T1 is used in only one innovation driver ID1 and does not have a large enough revenue potential and is hence better licensed.
- T2 and T3 are used by more than one innovation drivers and hence are candidates for licensing.

# 4 Discussions and conclusions

This paper identified the challenges faced by owners of technology innovations in deciding the best means of monetizing their technology innovations. It suggested that monetizing technology innovations is perhaps best carried out through a matching of properties of the technology innovations with value drivers or properties demanded by a business innovation. The paper defined TechVal matrix as a means of organizing the properties of technology innovations and value drivers. It described a method for determining the best means of monetizing technology innovations based on the extent of demand and the addressable market size. The paper also presented some examples of technology innovations that were originally created for one market being much more successful in other markets.

This method is perhaps the first to use a property based approach to plan the path to monetizing technology innovations. The method proposed is still work in progress and needs to be rigorously tested in real situations by those responsible for monetizing technology innovations. Feedback from such exercises will certainly help refine the method.

The proposed method is only a beginning in matching technology innovations' properties with the value drivers and there is room for further research. For example, the proposed method assumes that a technology innovation fully satisfies a value driver of a business innovation. This assumption may not indeed be true in many cases. One can design a method that can offer solutions where a technology innovation can only either partially or almost fully satisfy a value driver. A method that addresses partial matches is a research in progress and will be repeated in a future publication.

## References

Abernathy, W.J. and Utterback, J. (1978) "Patterns of industrial innovation," in Tushman, M.L., and Moore, W.L. Readings in the Management of Innovation, 97-108, Harper Collins, New York.

Christensen, C.M. (2003) Innovator's Dilemma, Harvard Business School Press, MA.

Inventors.about.com, http://inventors.about.com/library/inventors/blsteamengine.htm

Kelly, P. and Kranzber, M. (eds) (1978) Technological Innovation: A Critical Review of Current Knowledge, San Francisco Press, San Francisco, CA.

Kim, C. and Mauborgne, R. (2005) Blue Ocean Strategy, Harvard Business School Press.

La Salle, R. (2002) Think New, Rudders RLS Pty Ltd., Victoria, Australia.

La Salle, R. (2006) Think Next, Rudders RLS Pty Ltd., Victoria, Australia.

Leifer, R., Colarelli O'Connor, G., Peters, L.S., (2000) Radical Innovation, Harvard Business School Press, Boston, MA.

Narasimhalu, A.D. (2005) “Innovation Cube: Triggers, Drivers and Enablers for Successful Innovations”, Proceedings of the Annual ISPIM conference, Porto, Portugal.

Rothwell, R. (1992) “Successful Industrial Innovation: critical factors for the 1990s”, R&D Management, Vol. 22, No. 3, 221-239.

Schumpeter, J.A. (1934) The Theory of Economic Development, Harvard University Press, Boston, MA.

Schumpeter, J.A. (1939) Business Cycles, McGraw-Hill, New York.

Schumpeter, J.A. (1942) Capitalism, Socialism and Democracy, Allen & Unwin, London

Trott, P. (1998) “Growing businesses by generating genuine business opportunities”, Journal of Applied Management Studies, Vol. 7, No. 4, 211-22.

Utterback, J. (1994) Mastering the Dynamics of Innovation, Harvard Business School Press, Boston, MA.

Van de Ven, A.H. (1999) The Innovation Journey, Oxford University Press, New York.

Von Hippel, E. (1978) “Users as Innovators”, Technology Review, Vol. 80, No. 3, 30-34.

# The Importance of Social Innovation in the Dutch Manufacturing Industry:
## Innovation as a Joint Effort Between Research, Education And Business

**Saskia Harkema**
The Hague University of Applied Sciences, The Netherlands

**Editorial commentary**

In this study, Harkema reflects on a collaborative partnership initiative being conducted at The Hague University of Applied Sciences to help transform ideas into commercially viable products. The collaboration involves the University and a number of manufacturing SMEs from the surrounding area. The partnership initiative was established with dual aims: first, to expose both students and academics to the commercial sector and to challenges associated with commercialisation processes; and second to help increase the innovative capabilities and problem-solving skills of small manufacturing firms.

A detailed explanation is provided of the model being adopted at The Hague University of Applied Sciences, and of the contribution it has made to creating successful linkages between education and enterprise.

Some points for discussion and learning arising from this case include:

- Where should governments invest in innovation: in industry or academia, or both?
- The factors that contribute to the success of partnerships between universities and SMEs.
- The learning value for students and academics of being exposed to the working of SMEs and their challenges to innovate successfully.

---

**Abstract**: Innovation and entrepreneurship are so important in the realm of national economies because they hold the key to the continuity and growth of companies (e.g. Hage, 1999; Cooper, 1987; Van de Ven, 2007) and economic growth within a country. It is therefore obvious that national governments are spending a lot of money to enable and improve innovation management and entrepreneurial behaviour within organizations. This is also the case in The Netherlands who have voiced the ambition to become one of Europe's frontrunners in innovation and entrepreneurship by 2010.

Given the importance of both phenomena, a lot of effort is geared towards implementing entrepreneurship and innovation as topics within curricula at universities. The objective is to stimulate innovative and entrepreneurial attitude and behaviour, and motivate students to start their own businesses and develop knowledge and competencies about how to do so. In The Netherlands there are several initiatives that purposefully try to embed entrepreneurship and innovation as a subject within a diversity of professions varying from industrial management to behavioural sciences and business students. Despite these efforts, effects so far are meagre and improvements need to be made. In addition, The Netherlands are still lagging behind when it comes to their position in comparison to European counterparts in meeting the so-called Lisbon objectives. The Ministries of Economic Affairs and Education in The Netherlands have joined forces on this issue and given top priority to the development of so-called innovation program s centred on partnership and co-operation between educational institutes and the practical world of entrepreneurs. These partnerships should involve universities (education), companies (preferably SME's) and industrial associations (business) and representatives from governmental organizations (community) and should be geared towards: the development of sustainable networks, a contribution to regional economic growth within sectors, the development of learning communities in which best practices are shared, knowledge circulates and knowledge is created through applied research and last but not least sustainable relations are developed between universities and the business community.

Universities of Applied Sciences (UAS) have a special role in responding to and addressing these changes. In comparison to traditional universities the education offered at UAS is more multidisciplinary and oriented to the solution of practical problems. The development of problem-solving capabilities has subsequently become a central dimension of professional education according to Weert & Soo (2009). Against this background professional education at the highest level, further training of professionals in the workforce and greater enterprise-academic collaboration are seen as prerequisites to meet the demands of the future. The function of research and development must be aligned with these demands, reason why in the knowledge landscape UAS have to develop into teaching and research institutes.

Within the Centre for Innovation and Entrepreneurship (CI&E) at The Hague University of Applied Sciences in The Netherlands we have taken the initiative to develop an innovation program for the manufacturing industry to help them make the leap from idea to implementation, while simultaneously professionalizing students and teachers. This industry holds a very important position within the Dutch economy but simultaneously is under heavy pressure as a consequence of fierce international competition and outsourcing of production to cheap labour markets. These threats are countered by opportunities such as the application of new materials and processes, the introduction of new products and services and strategic repositioning in the supply chain. Given the economic importance of the sector the Ministry of Economic Affairs has allocated funds to the innovation program developed by our CI&E.

In the first year of the program which started in September 2007, 22 companies have participated. The leading questions in our research are: how do SMEs in this sector innovate and is the model we have developed which simultaneously tries to educate students, whilst stimulating innovation at company level and in second instance contributing to regional development, a viable model? In this paper the preliminary findings of our research so far are presented.

**Keywords**: Innovation, Research, Education, Industry, The Netherlands.

# 1 Introduction

In this paper we present a model which links entrepreneurship education with innovation management in the Dutch manufacturing industry (Van der Woude and Harkema, 2008). The model was developed at the Centre of Innovation & Entrepreneurship at The Hague University of Applied Sciences as part of an innovation program purposefully developed for this sector. The aim of the model is threefold. On the one hand the objective is to create an architecture in which students, lecturers and companies can share knowledge, learn and work towards specified goals. On the other hand the objective of the program is to contribute to the innovative power of participating sme's and development of problem-solving skills. And last but not least the program aims to contribute to the professionalization of both students and lecturers, i.e. teach them research skills, develop a reflective attitude towards the knowledge they have acquired during their study and develop an entrepreneurial and innovative behaviour. In this paper we describe the program and present the model and the research that has been carried out so far and in which 22 companies have participated. It is an exploratory research which aims to unfold a number of issues besides the three mentioned above: how do companies in this sector innovate and what is the role of the entrepreneur in the process, and to what extent are innovation models used by large companies useful to organize the innovation process in these companies? The main purpose is to explore what the importance is of social innovation in these companies as opposed to technological innovation. Social innovation refers to the renewal of organizational processes in such a way that it contributes to the quality of work (Volberda, 2006).

The paper is built up as follows. First a brief overview is given of the specific problem in this sector in The Netherlands. In addition the role of governmental policy and organizations in stimulating innovation and entrepreneurship education to address this and similar problems are discussed. In the next section the role and importance of entrepreneurship education is highlighted in addition to an overview of research into the relationship between education and economic growth. Basically this assumed relationship forms the backbone of our research and governmental policy since research points out that here is a positive correlation between education and successful entrepreneurship (Cardia et al, 2008), reason for governments to invest in education and stimulate the development of programs as the one described here. The innovation program is the topic of the next section where we elaborate on the content of the program, the set-up and the model developed by us. Subsequently we report on the research carried out by us among the first batch of 22 companies that participated in the program. We finalize with preliminary conclusions on what the research tells us about the model and the extent to which it contributes to regional development and improvement of the innovative power of individual companies i.e. social innovation. Our main focus of attention is however the extent to which this model achieves in realizing a linkage between education and enterprises and in that process contribute to the simultaneous professionalization of students, lecturers and teachers.

## 2 Governmental policies: Stimulation of innovation and entrepreneurship

National governments are investing heavily in entrepreneurship and innovation. Primarily for economic reasons and in second instance for social reasons. From an economic perspective innovation and entrepreneurship contribute towards national and organizational growth and the creation of new jobs. Reason why governments and local municipalities invest heavily in the stimulation of new-start-ups and technological innovation. The influence of entrepreneurship on economic growth can be measured in direct and indirect effects (Van Stel, 2009). In general one can say that the contribution of entrepreneurship towards economic growth is positive. In line with what was agreed among the 14 European countries that participated in the Lisbon meeting in 2000, entrepreneurship is however not merely understood as an activity geared towards the start-up of new firms. The behavioural aspect of entrepreneurship is regarded as important and a necessary condition for successful and innovative start-ups.

Based on a European vision the European Union has defined entrepreneurship as follows:

"Entrepreneurship is the ability to put ideas into action. It encompasses creativity, innovation and risk taking, as well as the ability to plan and conduct projects to realize objectives. An entrepreneurial attitude helps everybody in daily life, at home as well as in the society. It helps employees to become aware of their work environment and to grab chances as they occur, and it is the basis for more specific competencies and knowledge that entrepreneurs need for social and economic activities."

According to this definition, entrepreneurship is much more than a certain work attitude or competence, and as relevant for the employee as well as for the entrepreneur. Entrepreneurialism contributes to the quality of work and life in general.

## 2.1 The Dutch agenda on entrepreneurship

The European vision can be tailored to the Dutch situation, by observing closely the national environment. The following picture emerges from recent figures from the Dutch Central Bureau of Statistics (2007):

- Innovation and entrepreneurship are the greatest bottlenecks in the Netherlands
- The lack of innovation is caused by a lack of entrepreneurship.
- The Netherlands are successful in developing technological knowledge

Recent figures show (Nootenboom et al, 2008) that the economic arena in The Netherlands is characterized by a majority of sme's and a few multinationals like Shell, Philips, Unilever, etc. According to these figures there are around 770.000 sme's in different categories: knowledge intensive, innovative developers, followers and knowledge extensive companies. The National Council of Science and Technology has formulated an advice especially in relation to the innovation of developers and followers which amount to more than 200.000 companies. In general the Dutch Innovation system is regarded as inadequate to deal with the future challenges. Therefore innovation needs to be stimulated especially by recombination of existing knowledge. How? By stimulating demand as well as supply of knowledge and creation of regional networks between companies and researchers. Universities of Applied Sciences play an important role to realize these objectives. Rather than instructing students in academic and scientific subjects, UAS emphasize competencies oriented to professional practice. An entrepreneurial attitude and behavior are seen as necessary conditions to meet the requirements in the future field of work. Additionally students of UAS are expected to develop research skills and become critical consumers of academic literature (HOP 7, 2009).

The Ministry of Education, Science and Culture together with the Dutch Association of Employers and the HBO-Raad have set up a foundation – Foundation Innovation Alliance – in which also participate the national institute for applied sciences (TNO), a public intermediary for innovation consultancy (Syntens) and the Telematica Institute – a top institute for research collaboration on ICT. One of the main initiatives of the foundation are the so-called RAAK program s which stand for Regional Attention and Action for Knowledge Circulation.

# 3 The Dutch manufacturing industry

The Dutch manufacturing industry forms a factor of great economic importance. The sector offers jobs to more than 1 million people and indirectly offers employment to the same number. Additionally industrial companies account for 70% of Dutch export and nearly ¾ of R&D expenses. The manufacturing industry has a lot to offer to the Dutch economy, but recently has been confronted with some major changes which have had their effect. On the basis of a survey carried out among 1400 companies in 2006 a description is given of the expected competitive position of the sector in the European arena. The main conclusion is that the sector needs the support of the government while companies in the sector need to gear their activities towards the improvement of their competitive position. How? By investing in co-operation with schools and educational institutes, by focussing on sustainability and through innovation and co-operation.

In recent years the sector has been confronted with a trend towards outsourcing of manufacturing activities to cheap-labour markets and lack of investment in product and process innovation. Research points out that sme's in general are less innovative than their large counterparts (Jong, 2006). If we look at the structure of the Dutch economy we see that the largest portion (with a number 770.000 nearly 98%) is formed by sme's while only a mere 2% are large companies, i.e. multinationals like Shell, Unilever, Philips.

Against this background Dutch governmental policy has been formulated to counter the threat resulting form a lack in investment in R&D and outsourcing of activities. In the next section we discuss the measures that the Dutch government has taken to counter this trend one of them being support of innovation and entrepreneurship in education in partnership with economic sectors.

# 4 An Innovation Program to accelerate innovation

Against the sketched background a program was developed for the manufacturing industry with the aim to accelerate innovation within at least 50 companies. The program is subsidized by the Stichting Innovatie Alliantie -

a foundation set up by the Dutch Ministry of Educational Affairs. The objective of the Foundation is to stimulate the development of innovation programs in partnership between universities, companies and intermediary supporting organizations, i.e. between education, research and the companies in a certain region. The program was developed at the Center of Innovation & Entrepreneurship in partnership with associations of which companies are members and an organization which acts as an agent for the Ministry of Economic Affairs (Syntens) in supporting companies to improve their innovative capacity. It is a two-year program for which a model was developed as depicted below:

The central issue of the program is that SMEs within this sector encounter difficulty in making the leap from strategy to implementation. They basically know what to do, but not how to do it. Therefore, the key question in the program is how companies can define innovation strategies in line with their internal organization and what do they need to do to implement these strategies. Within a two year period the program aims to help 50 companies through a tailor-made educational and research program in which students, lecturers, and business people participate and thus create a learning community evolving into a sustainable network after those two years. The program consists of a company diagnosis by means of a purpose-built software tool, expert meetings and master classes on topics specified by the team, consisting of lecturers, students, business people and researchers. Topics for these meetings range from leadership and change; lean manufacturing to human resources management. Finally, companies are presented with their individual definition of interventions to purposefully work towards specified goals and objectives. In the process from diagnosis and analysis, through action towards intervention; research is carried out to distil lessons from practice, enabling us to define competencies that people need to acquire to improve strategic innovation processes in future, and to describe conditions that should be met to align the internal organization with external developments and ensuing strategic choices from those developments.

The program is split up in 4 periods of 20 weeks in which groups of students and lecturers work together with 10 to 15 companies. The 20 week-period consists of a phase of analysis followed by a period of implementation and defining of actions. During that period so-called network meetings are organized around themes relevant for the companies. Parallel to the process research is carried out. This cycle is repeated 4 times.

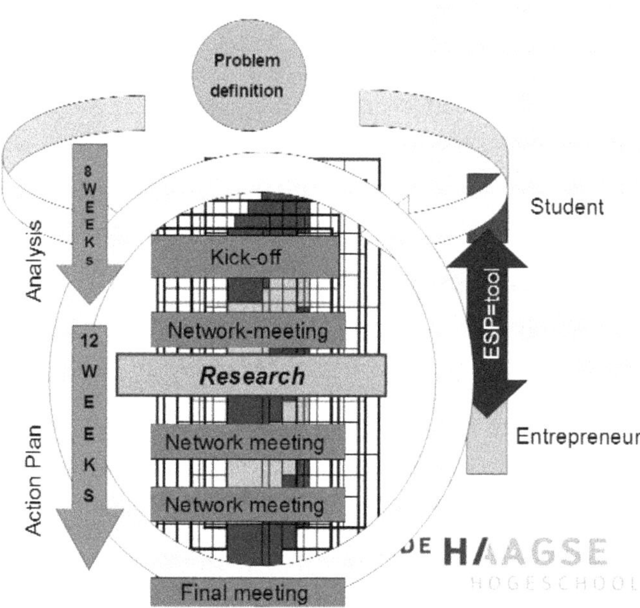


**Figure 1:** Repeating Cycle

# 5 Research design

Parallel to the program which is aimed at supporting the companies in development of professional skills and knowledge how to improve their innovative capacity, research is carried out to gain insight into the bottlenecks and problems encountered along the road. As mentioned in the introduction the aim of the program is threefold:

- On the one hand the objective is to create an architecture in which students, lecturers and companies can share knowledge, learn and work towards specified goals.

- On the other hand the objective of the program is to contribute to the innovative power of participating sme's and development of problem-solving skills.
- And last but not least the program aims to contribute to the professionalization of both students and lecturers, i.e. teach them research skills, develop a reflective attitude towards the knowledge they have acquired during their study and develop an entrepreneurial and innovative behaviour.

It is an exploratory research which aims to unfold a number of issues besides the three mentioned above: how do companies in this sector innovate and what is the role of the entrepreneur in the process, and to what extent are innovation models used by large companies useful to organize the innovation process in these companies? As a framework the model of Tidd, Bessant & Pavitt (2005) was used. In this model the innovation management process is defined as a series of phases companies have to go trough (scanning, strategy, resourcing, implementation and learning) and routines companies have to develop in order to manage innovation successfully. Our interest was to explore whether companies work in a similar way as described in the model.

The research group consisted of 22 companies in the manufacturing industry in the province of South-Holland. 10 companies have less than 10 employees and 11 companies between 11 and 22 employees. Turnover figures vary between Euro 500.000 to Euro 10. mln.

The research was carried out through a baseline measurement at the start of the program, a scan of the company with the help of a so-called Electronic Strategic Planning (ESP) tool and by in-depth interviews with all the directors/owners of the companies. The interviews were semi-structured.

The Electronic Strategic Planning (ESP) Tool is a software tool which in a first step expects a formulation of the mission and vision of the company and identification of product/market combinations followed by an external analysis. The last part of the scan is related to the financial situation of the company. This forms the input for a SWOT analysis after which a score is given to the company on 6 profiles based on 3 dimensions: cost-focus, customer-focus or product differentiation. An entrepreneur then has to make a choice out of the 6 profiles from which an action plan follows. The research also looked into the role and usefulness of the instrument as a tool for strategic choices.

Companies were primarily evaluated to the extent that through participation of the program behaviour changed leading to new product innovation or process innovation or organizational renewal.

The assessment of the students took place via the reports in which they on the one hand had to write an action plan for the company where they had been based during the 20 week period of their practice. The report also assessed students on research skills, their ability to apply knowledge in a practical context and their capacity to reflect on that knowledge and offset it against existing theories.

Lecturers were assessed on two aspects: the extent to which they adapted their curricula on the basis of the experience and knowledge gained and the broadening of their knowledge base as a result of the supervision of students and their own involvement with the companies.

## 6 Research findings

In the measurement entrepreneurs were asked to indicate what the barriers for innovation were on a 5 point Likert-scale: never, sometimes, once in a while, often, always. This first measurement shows that the companies certainly innovate, but that most innovations are incremental. Drivers for innovation are competitive pressure, cost reduction and technology push.

**Table 1:** Barriers for innovation Source: Baseline measurement (Sept 2007 & March 2008)

| **Marketing** | **18** | **Difficulty in finding partners for innovation** | 10 |
|---|---|---|---|
| **Lack of suffciently qualified personnel** | **19** | Interest of customer uncertain | 14 |
| **Large economical and financial risks** | **17** | Bad exchange of knowledge (internal and external) | 13 |
| **Lack of knowledge on market developments** | **19** | Administrative hassle | 7 |
| **High direct costs for innovation** | 18 | Rules and laws | 9 |

| **Marketing** | **18** | **Difficulty in finding partners for innovation** | 10 |
|---|---|---|---|
| Insufficient knowledge about technological developments | 16 | ICT support | 11 |
| Availability of means and money to innovate | 17 | Lack of flexibility in existing procedures and rules | 7 |
| Lack of insight in the own need for knowledge | 15 | IP issues | 9 |

The first measurement on barriers of innovation according to the entrepreneurs brought the following to light (Van der Woude & Harkema, 2008, p.26):

*After the baseline measurement every owner/director was interviewed to establish how the innovation process, was managed, the vision on innovation, perception on the innovative capacity of the company and opinion about the ESP tool.*

### 6.1.1 Scanning

The research shows that when it comes to identifying opportunities, making a selection and translating these into a commercially viable solution, little use is made of external information. Lack of time is often given as a reason for not scanning the environment systematically, in combination with not knowing what choices to make.

### 6.1.2 Strategy and selection

Though entrepreneurs are motivated to make strategic choices and define strategies in line with the own organizational capabilities and competencies, they lack tangible and objective means to so. A tool like the ESP tool is regarded as an objective tool which on the one hand gives insight in the company and can be used to give focus and direction to the company. The research shows that entrepreneurs have a hard time in relying solely on the outcome of such an instrument. It is often one of the reasons that a consultant is hired in order to support an entrepreneur in his decision-making process. Not seldom an entrepreneur tends to follow his own plan and strategic choices are the result of what the owner/director of a small company should be done.

### *6.1.3 Implementation and learning*

In the implementation phase the step from strategy to action is made. It is then that process management gains in importance and routines can contribute towards the successfulness of an innovation. A structured approach is more an exception than a rule in the companies that participated in the program. Also in this phase the owner/director plays a key role. Learning and definitions of lessons is therefore not common practice because the owner/director does not see the added value, or the need to do so.

In summary, what the research brought to light is that within these companies innovation is not approached in a structured and systematic way. In general product innovations are primarily incremental and not the result of a proactive attitude towards innovation. These companies react to the market instead of leading the market. The use of the ESP tool brought to the surface that there is overall insufficient knowledge within the company about the market and market development, nor up-to-date knowledge of technological developments. Despite the fact that the directors interviewed voiced a need to have an objective measure to define their strategy, their own opinion and thoughts about what that strategy should be are dominant. They did express that working this way in partnership with students, lecturers/researchers and other companies changed their way of thinking about the possibilities of co-operation for innovation.

# 7 Conclusions

One of the objectives of the innovation program described here is to create an architecture in which students lecturers and companies can share knowledge, learn and work towards specified goals. The research shows that through the model we have created a platform in which students, lecturers and companies work together. Companies indicate the biggest added value of the platform created is that:

- an in-depth analysis is made of their company
- students in a 20-week period can work fulltime with them in defining a strategy for the company
- a network is created with other companies and thus opportunities for partnerships emerge and are created and come to life.
- knowledge-sharing is stimulated and knowledge is gained through the network meetings

- issues are addressed in the network meetings for which they in their day-to-day business have not time to think about nor reflect on
- new approaches can be tried out and best-practices can be developed from which other companies can benefit

So far the research shows that creating such a platform has positive effects in terms of networking, knowledge sharing, co-operation and partnerships.

One of the other objectives of the program is to contribute to the innovative power of participating sme's and development of problem-solving skills. On the basis of the first findings it is difficult to assert whether direct contributions have been made to the innovative capacity and entrepreneurialism of people and organization. Some companies acknowledged the fact that through this program and way of working they adapted their own way of working and routines, but the findings are as yet inconclusive. More research needs to be done to establish whether the changes in behaviour are sustainable and whether a direct link can be made between this changed behaviour and an increased investment in product and process innovation. The program runs till September 2009 and we will continue evaluating and monitoring the whole group of 50 companies at least one more year after finalization of the program.

And last but not least the final objective of the program is to contribute to the professionalization of both students and lecturers, i.e. teach them research skills, develop a reflective attitude towards the knowledge they have acquired during their study and develop an entrepreneurial and innovative behaviour. The results so far show that through this program students participate in a company in a different way than their usual practice. In this case they act as a consultant with the owner/director as their counterpart. The result is that they gain a broad view of all the problems one can encounter within a small and medium sized company and – more importantly – the way a director approaches and tackles these problems. In terms of educating entrepreneurship they experience directly what it means to make choices on the basis of opportunities and define a strategy that fits with the company while being adaptive to the changes the external world requires. The perception students have of SMEs transforms drastically as a consequence of this experience and also their perception of the role an entrepreneur plays in the management of the company and the decision-making process.

The question is however: to what extent are students of UAS able to approach a problem from a different angle than they are used to - to solve that problem directly – namely to first analyse the situation, reflect on what is occurring and define alternative strategies an entrepreneur could pursue to tackle the problem at hand. And in so doing use different knowledge sources instead of one model as a basis to define those strategies and alternatives? Basically this lies at the heart of the idea that students from UAS not merely have to apply knowledge and implement, but first reflect on the applied knowledge and thus become critical consumers of academic knowledge. Entrepreneurial and innovative behaviour in this sense refers to a pro-active attitude towards what they are being taught and how it is taught and develop a critical and reflective approach towards the extent to which knowledge can be applied in a certain context. In this way the gap between education and practice can be bridged, because knowledge which has a proven record in practice – i.e. is evidence-based – can then be fed back into curricula. This is what knowledge circulation implies.

The research shows that a key role in this process of professionalization is played by the lecturers that supervise the students. They act as role models and are the source of inspiration and knowledge. This is one of the conclusions from the research. While at the start of the program we focussed on the student as the primary target group in the process of professionalization, in the course of the project we came the conclusion that first of all lecturers need to have the necessary skills. This proves to be a bottleneck. Teachers and lecturers at UAS are not used nor trained to do research and lack the skills and experience to do so. Since they act as supervisor for students and their results they need to acquire those skills and competencies first. This is part of the program, but our findings cannot confirm whether there is a positive direct effect. What does occur is that lecturers feed back their experiences directly into their own curricula and as such contribute to the actualization of knowledge and practice-based learning. From the perspective of developing entrepreneurial skills – i.e. entrepreneurialism - and innovative behaviour we can conclude that the direct confrontation with companies and the participation in networks with companies has a positive effect on their own knowledge and mind-set about what entrepreneurship means and requires.

## 8 Future research

This is an account of an exploratory research which reveals the results after one year of doing research. The program will run till September 2009 and the aim is to have at least 50 companies follow the trajectory from analysis to implementation and action. The evaluation and monitoring of the company will continue after September 2009. At present 8 more companies have gone through the program and this confirms our findings so far. Future research will concentrate on the questions defined in this paper and additionally focus our attention on a) development of research and problem-solving skills and innovative and entrepreneurial behaviour among students and lecturers and b) relation between changed behaviour of entrepreneurs and investment in product and process innovation and c) organization of the innovation process.

## References

Bharadwaj, S. and Menon, A. (2000) Making innovation happen in organizations: individual creativity mechanisms organizational creativity mechanisms or both?, in Journal of product innovation management . 17(3) pp 424-434.

Cardia F., Van Praag M. (2008) "Onderwijs en ondernemerschap in Nederland", Amsterdam center for Entrepreneurship: Amsterdam.

Cooper R.J. (1987) "Winning at new products", Kogan Page: London.

Dom F., Harkema S.J.M. and Jousma H. (2007) Sectorraamwerk voor het HO: "Ondernemendheid en Ondernemerschap in het Hoger Onderwijs: voor iedereen die het verschil wil maken", Senter Novem: Den Haag.

European Growth and Jobs Monitor (2008), Indicators for success in the knowledge economy, www.lisboncouncil,net.

Gibb A.A. (2005) "Towards the Entrepreneurail University", Policy paper # 3, National Council for Graduate Entrepreneurship: Birmingham.

Hage J.T. (1999) "Organisational innovation and organisational change", Annual Review Sociology, Vol 25(7): 597-622.

Hermann K, Hannon P. and Cox J. (2008) "Developing Entrepreneurial Graduates: Putting Entrepreneurship at the centre of higher education", NESTA NCGE and CIHE: Birmingham.

Hoogeschoolontwikkelingsplan HOP-7 2009-2013, The Hague University of Applied Sciences: The Hague.

Jong de, J.P.J. (2006) Innovatie in het MKB, EIM: Zoetermeer.

Mascitelli, R. (2000) From experience: harnessing tacit knowledge to achieve breakthrough innovation. Journal of product innovation management, 17(3): pp 179-193.

McDermott, C.M. and G.C. O'Connor (2002) Managing radical innovation: an overview of emergent strategy issues, in Journal of product innovation management pp. 424-438.

Nootenboom B. and Stam E. (2008) "Innovatie vernieuwd. Opening in viervoud" The Netherlands Scientific Council for Government Policy, Amsterdam University Press: Amsterdam.

Onstenk J. (2003) "Entrepreneurship and vocational education, European Educational research Journal, Vol 2(1) pp 74-84.

Schout H.J.; Muntjewerff E. and Harkema S.J.M. (2008) "Creating a pedagogical climate to stimulate entrepreneurial behaviour", proceedings of the 3rd European Conference on Entrepreneurship and Innovation, Academic Publishing Limited: Reading

Schumpeter J. (1934) The theory of economic development. An inquiry into profits, capital, credit, interest and the business cycle, Harvard University Press: Cambridge.

Sijde van der P., Mc Gowan P. and Kirby D. (2008) Guests editors' introduction to the special issue: The entrepreneurial spirit in higher education and academic entrepreneurship. Industry & Higher Education, Vol 22(1)3-8.

Slotman, R. (2007) "Stimulating regional Innovation in co-operation with SMEs" Berlin.

Soosay, C.A., An Empirical Study of Individual Competencies in Distribution Cetres to Enable Continuous Innovation. Creativity and Innovation Management, 2005. 14(3): pp 299-310.

Tidd J., Bessant J. and Pavitt K. (2005, 3rd edition) "Managing innovation, Integrating technological, market and organizational change", John Wiley and Sons: Chichester.

Van de Ven, A., Polley, D., Garud, R. and Venkataraman, S. (2007) "The innovation journey", Oxford University Press: New York.

Van der Woude H. and Harkema S.J.M. (2008) "Innoveren binnen handbereik". Ee exploratie naar MKB bedrijven in de maakindustrie, LuLu: The Hague.

Van Stel A. (2009) “Hoe beïnvloedt ondernemerschap economische groei”, Amsterdam Center for Entrepreneurship: Amsterdam.

Volberda, H.W., Van den Bosch, F.A.J. & Jansen, J.J.P. (2006), "Slim Managen & Innovatief Organiseren", Eiffel ism Het Financieele Dagblad, AWVN, De Unie & RSM Erasmus University.

Weert de, Geert & Soo, Maarja (2009) “Research at Universities of Applied Sciences” European project: Educating the new professional in the knowledge society, University of Twente.

# Intermediaries in the Management Process of Innovation: The Case of Danish and German SMEs

**Susanne Gretzinger[1], Holger Hinz[2] and Wenzel Matiaske[3]**
[1]University of Southern Denmark, Denmark
[2]University of Flensburg, Germany
[3]Helmut-Schmidt-University Hamburg, Germany

**Editorial commentary**

Gretzinger, Hinz and Matiaske discuss the role of consultancies as intermediaries in innovation processes in SMEs, in particular from the point of view of the acquisition and management of information. They consider the advantages and disadvantages of using external agencies, such as consultancies, noting for example, how there can be a difficult balance to strike between leveraging additional resource, and yet risking outflow of knowledge to competitors.

The study was conducted among SMEs in Denmark and Germany using survey-based techniques.

Some points for discussion and learning arising from this case include:

- How SMEs can best manage their relationships with key intermediaries, such as private or public-sector consultancies.

- The similarities and differences between private and public-sector consultancies with regard to SME business growth and innovation support.
- Approaches SMEs can adopt to reach an appropriate balance between leveraging additional resources from intermediaries and risking leakage of information and knowledge to potential competition.
- The purpose and scope of contractual agreements between SMEs and their intermediaries.

**Abstract**: Information is a critical resource in innovation processes. SMEs are therefore advised to draw on consulting in innovation processes, as they cannot ensure the necessary information flow internally due to lesser resources. From the strategic point of view, the involvement of intermediaries is accompanied by the risk of losing specific knowledge to the business environment. But the other way around: yo neglect the integration of consultancies could mean a deficit in the process of information management. Based on an empirical comparative study of Danish and German SMEs – Danish companies utilize public as well as private consulting services more often – determinants of the usage of business consultancies in innovation processes are explored. One important implication of the paper is, that the consulting system has difficulties in reaching smaller SMEs and should be modified to reach SMEs much better.

**Keywords**: SMEs, consultancies, innovation, network

# 1 The role of consultancies, customers and suppliers in the process of innovation

In organization theory it is agreed that consultancies, whether public or private, institutionally take on the role of an intermediary. Consultants communicate information on structure and strategy from one business to the next and thereby ensure the formation of relatively homogeneous organization popu-lations and the alignment of organizations' phenotype, respectively. Hannan/Freeman (1984), the principal agents of the population ecology approach, and DiMaggio/Powell (1983), from the perspective of the competing research program of neo-institutionalism, agree on that. Thus DiMaggio/Powell (1983: 151) write about the role of business consultancies: "Models [of organizations, the authors] maybe diffused unintentionally, indirectly through employee transfer or turnover, or explicitly by organizations such as consulting firms or industry trade associations. Even innovation can be accounted for by organizational modeling." With regard to innovation processes Wu et al. (2009) and Wolpert (2002) work out the dissemination of best practices via consultancies, whose role they correspondingly characterize as "innovation intermediation".

The risk of knowledge outflow to a competitor in the consulting process to the detriment of one's own business is, however, not as high as it might seem at first sight. Consultants do not practice industrial espionage in order to deliver blueprints specifically and synchronously from one company to another. Although consultants are chosen because of their experience with problems similar to the client's, this experience will, however, hardly solve the client's problem as exactly as the proverbial missing link. The experience brought in by the consultant from – usually previous - other consulting processes first needs to be mutually interpreted, understood and adapted. So even if organization theory proves ignorant towards the clause of client protection in contracts with consulting firms and fuels mistrust, practitioners have good reasons – due to the asynchrony of information transfer, as well as the lack of specificity, which goes along with the need for interpretation of the practical consulting knowledge – to trust consulting firms from time to time in the case of innovation processes. They also count on it that the potential gain in experience is opposed to a merely small risk of the unwanted transfer of know-how (Glückler/Armbrüster, 2006; Wu et. al., 2009).

However, companies trust consultants rather than their remaining organizational environment: apart from those spectacular cases where hiring a consulting firm serves the legitimization and defence of decisions that were already taken and where it needs to be announced that a certain consulting firm is engaged in-house, only little is disclosed about the utilization and benefit of consulting. Perhaps companies do not want to convey the fatal signal of their weakness. It is therefore not surprising that so far only few studies are available on the particularities of innovation consulting.

A central empirical finding from Glückler and Armbrüster (2003: 289-290) states that the involvement of consultancies is accompanied by a high level of uncertainty. This uncertainty results from a lack of sanction mechanisms on the part of the client, in order to defend themselves against the outflow of innovation knowledge. Also Wu et al. (2009: 3) state, based on in-depth interviews, that in an innovation consulting process an outflow of knowledge from the enterprise receiving advisory services can be expected. While these authors share also the general scepticism of organization theory towards consulting firms as intermediaries, Hislop (2002) arrives at a more differentiated assessment in his theoretical analysis of the relationship structure between company and consulting firm. With reference to innovation consulting, on which we focus here, he writes: "Interactive innovation, however, involves disparate social communities, which can have very different systems of meaning. Relying on embedded client-consultant relations, at least to some extent, appears to provide a way of lessening the difficulties of the knowledge sharing that is required in such interactive innovation process" (Hislop, 2002: 669). According to that, the above mentioned need for interpretation of the knowledge that is transferred in innovation consulting processes is mainly based on the relative closeness of social circles, between which consultancies build their bridges – to put it in network-analytical terminology. But even if consulting firms succeed in creating trust to their clients for these reasons – Hislop (2002: 665) talks about "swift trust" – a risk of knowledge outflow remains on the part of the clients: it is the consultants' business to collect and process practical knowledge obtained in consulting processes.

Summarizing these thoughts and findings of empirical studies, it can be noted that the consultancies' task of processing general information and practical knowledge from other consulting processes implies the risk of diffusion of company-specific knowledge for the clients in innovation consulting. Hence, consulting firms build bridges between information pools or are – in the previously introduced network-analytical terminology – weak ties from their clients' point of view. Apart from consultants, direct business partners, i.e. customers and suppliers, are considered important partners in the innovation process (Brockhoff 2003, von Hippel 1978). Companies take up customer wishes or supplier information and use them as a starting point for their product or process innovations. In this regard there are generally two options: the companies carry out the process alone or they cooperate with their customers or suppliers in the process of the innovation. The first constellation is not relevant for further analysis. Companies that innovate based on information from the business environment do risk an indirect outflow of information via e.g. staff changing to competitors, but they are not in danger of losing know-how that is relevant for the innovation in the relationship with customers or suppliers. However, this risk exists if cooperative relationships are entered in the innovation process.

Independent of the legal arrangements in these cooperative relationships, i.e. from merely implicit or formal contracts up to joint ventures, they are a constellation of mutually specific investments (Williamson 1985) or a combination of resources (Coleman 1974), which imposes the risk of loss on both parties. Therefore, customers or suppliers, as well as the innovating focal business, have in such relationships an interest in shielding off third parties from the innovation process (Afuah/Bahram 1995, 75; Reichwald/Piller 2005, 9). Following this argumentation, we hereafter consider direct contacts in the business environment as strong ties in the terminology of network analysis.

Another interesting aspect regarding the involvement of consultancies and/or customers and suppliers in the innovation process are the differences in the international comparison. There is evidence that weak ties are used far more in Scandinavian countries than in other European countries (Poulfelt et. al. 1994, Brodbeck et. al. 2000). But "weak ties" and "strong ties" alone are not the solely decisive factor for innovation success. Initially they are rather part of the relevant resources which are used. In order to have a reference point as to which role "weak" and "strong ties" play in the total innovation process, the next section looks generally at the management of resources in the innovation process.

## 2 Management and cooperation in the process of innovation

Relationships embedded in "strong ties" or "weak ties" are supporting the focal enterprise with knowledge, which is a central variable in the process of creative destruction and implementation of new combinations of production factors, as Schumpeter described the process of innovation (Schum-peter 2006). Schumpeter's elements in the definition of the innovation process clearly show that the relationship to intermediaries is not just important because they support the focal enterprise in the process of recombining resources. How to prevent the loss of strategic information and therewith of the competitive edge has to be discussed through the lens of strategic resource management.

Strategic management refers to a number of central theoretical frames to discuss how to foster the strategic position of the enterprises best. In this study the reference point is the research dependence approach (RDA) (Pfeffer/Salancik 1978), which seems to be particularly suitable for a number of reasons. Not only is the RDA considered theoretically well developed and empirically sound (Nienhüser 2008), but it is also specialized in the question of external relations of organizations as in our case to consultancies, customers and suppliers in the process of innovation. Following the criticism of the contingency approach, which has long dominated organizational theory, Pfeffer/Salancik (1978, see also Aldrich/Pfeffer, 1976) fall back on a power-theoretic argument (Emerson 1962) in order to clarify which situational determinants govern the behavior of organizations. With this theoretical foundation they provide a meta-criterion that limits the arbitrariness of situational influencing factors and explains why the environment has an influence: The resource dependency of the organization is the basis of external exertion of influence. As opposed to other resource-oriented approaches, resources are here defined not only as input but also as output factors, i.e. the access to pre-product markets can be considered as a resource, just like the one to the final sales market.

External control can be exercised by those actors that control resources which are significant for the organization's effectiveness. The level of the organization's demand determines how powerful the partner is: the greater the interest of the focal organization in resources that are under the control of an external actor, the greater the power and also the influence of just this external player on the focal organization. This argument entails, furthermore, that the better the external actor manages to monopolize the interesting resources, the more influence he can exert. Conversely, the more difficult it is for the focal organization to obtain the interesting resources outside the relation to the external actor, the greater his power in the focal organization. It is particularly useful for the influence on the organization if the external player controls resources that are vital for the focal organization. In this case, Pfeffer/Salancik (1978) talk about critical resources.

A suitable frame of reference for assessing the interests of businesses in an innovation process is in our context the relational view of the firm (RV) (Dyer/Singh 1998). In extension of the better known resource based view of the firm (RBV) (Wernerfelt 1984), which focuses on individual businesses and their core competencies, the RV identifies the relevance of networks for the companies' resources and for generating a competitive advantage. Just like the RBV, the RV is so far predominantly phenomenologically or normatively oriented (Duscheck 2004; Freiling 2008). However, the descriptive integration of business networks, competitive markets and core competencies of the individual businesses is here sufficient to derive specific constellations of interests. To explain these we refer back to the power-theoretic argumentation of the RDA. To sum up: Trust and control over critical resources (core-competences) in the process of innovation are the main impact factors explaining the usage of weak versus strong ties. Against this background we state the following hypothesis:

***Hypothesis 1***: *The better a business is equipped with resources, the more likely it is that consulting services are used in the innovation process.*

***Hypothesis 2***: *The better the contractual agreement of the consulting service, the more likely it is that consulting will be utilized in the innovation process.*

***Hypothesis 3***: *The stronger the trust in the consulting system, the more likely it is that consulting will be utilized in the innovation process.*

***Hypothesis 4***: *The more important the innovations for the business, the less likely it is that consulting will be utilized in the innovation process.*

# 3 Methods

## 3.1 Data base and measures

The data set of this study is based on a postal (Denmark) and a telephone (Germany) survey on the innovation behavior of SMEs and on the utilization of the consulting system in both countries. In both countries two surveys were conducted: one in businesses, the other in public and private organiza-tions offering innovation consulting services. According to the focus of this study only the business data are used here.

The population of SMEs was limited by the target criteria location, size and industry. On the Danish side, businesses from Jutland and Funen were included, while it was SMEs from the federal states of Mecklenburg-Western Pomerania, Hamburg and Schleswig-Holstein in northern Germany. Businesses from the population do not employ less than 5 and not more than 500 members of staff and are from the goods-producing industry. Both partial surveys were carried out based on random samples. The return rate of the postal survey in western Denmark was roughly 12%. In Germany, approximately 31% of the interviews with businesses from northern Germany could be used. Only members of executive management were interviewed.

Table 1 lists the operationalizations of the variables that were used in the hypotheses. In the survey we asked in detail about cooperation in the innovation process. One series of questions dealt in general with the cooperation, the last innovation process in the past three years being the anchor point. Two other series asked in more detail about the last successful resp. unsuccessful innovation in the time period. Strong ties with cooperation partners in the innovation process are operationalized into relations to customers and suppliers.

**Table 1:** Description of Measures used in the Study

| Name of Variable | Operationalisation |
|---|---|
| "strong tie" | Cooperation with customers and suppliers in the innovation process |
| "weak tie" | Cooperation with public or private consultants in the innovation process |
| size | Number of employees |
| contract | 1) Was the partner subjected to specific test criteria before entering the cooperation? (yes/no)<br>2) Was a contractually binding agreement entered with the partner? (yes/no)<br>3) Was the partner subjected to specific test criteria after the completion of the cooperation? (yes/no) |
| trust | 1) Does your partner trust you? (4 fully, 1 not at all)<br>2) Do you trust your partner? (4 fully, 1 not at all) |
| Hauschildt-Schlaak index | Novelty of the innovation (Likert scale, 7 items, Cronbach's $\alpha$ = .91/.95) |

The tie-groups are usually mentioned jointly in the underlying multiple answer (r = .40). In total 52.8% of the businesses cooperated solely with customers and sup-pliers in the innovation process. Accordingly, cooperation with public or private consultants are subsumed as weak ties. Apart from a few exceptions, these businesses have both strong and weak ties. The two consulting categories correlate with r = .31. In total 34.3% of the enterprises did not enter any partnership in the last innovation process. With 14.7% Danish SMEs utilized weak ties slightly more often in the innovation process than the German SMEs, where public or private consultancies were used in only 11.3% of the cases.

The variables regarding the contractual agreement and trust in the partner in the innovation process are obtained through questions which describe the relation with the cooperation partner in more detail. We surveyed whether the partner was checked by the SME ex ante or ex post with specific criteria and whether there was an explicit contractual relationship with the partner in the innovation process. Furthermore, the trust relationship was reciprocally surveyed in self-assessment and the expected third-party assessment. This item set was subjected to a principal component analysis and was rotated orthogonally. As a result we receive two independent components, one of which depicts predominantly the contractual agreement, the other the trust relationship with the partner.

Another item set, which is known as the Hauschildt-Schlaak index, measures the degree of novelty of the innovation for the company. The items refer to the applied technology, channels of distribution, suppliers and production, the culture and structure of the organization and marketing costs (Hauschildt/Schlaak 2001). To determine an anchor point for this scale, the interviewees were first asked to describe in an open answer both the most successful and the least successful product innovation of the past three years. For each of these innovations, if available, we obtained the Hauschildt-Schlaak index. The reliability of the scale is remarkably high, with ⍰ = .91 for successful innovations and ⍰ = .95 for the unsuccessful innovations.

# 4 Results

Binary logit estimations are applied for the modeling. Target variable in all models is the utilization of weak ties in dummy coding

Corresponding to the hypotheses we developed, the models successively take on the variables for business size as proxy for resource equipment, the indices for contractual agreements and trust between the cooperation partners as well as the country in dummy coding (0 = DK, 1 = D). Extended models with additional control variables will not be reported here, as the variables of organization demography, which have so far been considered, do not lead to findings that are fundamentally different.

Table 2 reports the findings for the last innovation process in the past three years. The table shows the marginal effects, as those allow a direct interpretation of the direction and impact of effect.

**Table 2:** All enterprises, probability of utilization of "strong" vs. "weak" ties

| **predic-tors** | **(1) basic model** | **(2) + con-tract** | **(3) + trust** | **(4) + country** | **(5) + public consult** | **(6) + private consult** |
|---|---|---|---|---|---|---|
| size | 0.0467*** (0.0010) | 0.0487*** (0.0012) | 0.0489*** (0.0012) | 0.0511*** (0.0006) | 0.0255*** (0.0020) | 0.0380*** (0.0029) |
| contract | - | -0.0262 (0.2500) | -0.0263 (0.2480) | -0.0361 (0.1170) | -0.0144 (0.3010) | -0.0148 (0.4630) |
| trust | - | - | 0.0086 (0.7080) | 0.0127 (0.5760) | -0.0031 (0.8170) | -0.0002 (0.9910) |
| country | - | - | - | -0.1020** (0.0273) | -0.0109 (0.7000) | -0.1140*** (0.0036) |
| constant | -0.3990*** (0.0000) | 0.4050*** (0.00000) | -0.4060*** (0.0000) | -0.3540*** (0.0000) | -0.2490*** (0.0000) | -0.2910*** (0.0000) |
| n | 323 | 288 | 288 | 288 | 288 | 288 |
| LL | -154.24 | -135.75 | -135.75 | -133.29 | -72.61 | -113.71 |
| p | 0.0010 | 0.0015 | 0.0044 | 0.0013 | 0.0151 | 0.0021 |
| $R^2$ | 0.0323 | 0.0432 | 0.0437 | 0.0605 | 0.0657 | 0.0633 |
| Logit: Marginal effects for all SMEs with at least one innovation and cooperation partners. Probability p in brackets. | | | | | | |
| ***p<0.01, **p<0.05, *p<0.1 | | | | | | |

The signs of the marginal effects show the predictor's direction of effect, i.e. a positive sign indicates that the risk of the SME entering weak relations in the innovation process rises with a marginal increase of the independent variable. Along these lines it applies to the country dummy that the direction of effect needs to be interpreted with regard to the reference value – here Denmark. Therefore a negative sign implies that Danish SMEs will rather build up ties with consultancies than German businesses.

The results show that the model estimates are altogether significant throughout the analysis, but that explanatory contributions for the SMEs' decision behavior are, however, low. Pseudo $R^2$ values are between 3% and just above 6%. The variance explanation can hereby almost solely be referred back to the variables size and country. Compliant with the hypotheses, a better resource equipment of the business, represented here by business size, is accompanied by a greater usage of the consulting system. The variables of contractual agreement and trust in strong cooperation relations to customers and suppliers, which are significant from a theory perspective, do not influence the utilization of consulting in the innovation process according to these analyses. This holds also true if the consulting system is not analyzed as a single unit with regard to the target variable, but separately for public and private consultancies. In contrast, the differentiated analysis shows clearly that the significantly higher utilization of the consulting system in Denmark can be referred back to the more frequent involvement of private consultancies in the innovation process. In this respect, German SMEs are comparatively reserved, as already mentioned in the description of the data.

Similar results can be recorded for the analyses of the most and least successful innovation of the past three years, which is compiled in Tables 3 and 4. First of all, it should be noted that nearly all SMEs that generally reported an innovation in the relevant time period also had a successful innova-tion. In contrast, a less successful innovation can only be found in roughly half of the SMEs with innovations.

As before, we successively extend our base model by the variables size, contract, trust and the dummy for the differentiation of the countries. Contrary to Hypothesis 4, the relevance of the innovation process, measured with the Hauschildt-Schlaak index, does not change the usage pattern of the consulting system by SMEs. Only in a differentiated analysis do we find a significantly higher utilization of public consulting institutions in the case of less successful innovations. Spontaneously, this effect could be interpreted in such a way that in innovation processes which are important but where success is jeopardized, public consultancies are called in as friends in need. However, this single finding should not be overrated. For the country dummy, on the other hand, we find a familiar pattern. In contrast to German businesses, Danish SMEs utilize the consulting systems significantly more often. In the case of less successful innovations this only holds true for private consultancies, though, and not anymore for the consulting system in general.

**Table 3:** (Enterprises with successful innovation), probability of utilization of "strong" vs. "weak" ties

| **predictors** | **(1) basic model** | **(2) + contract** | **(3) + trust** | **(4) + country** | **(5) + public consult** | **(6) + private consult** |
|---|---|---|---|---|---|---|
| size | 0.0445*** (0.0043) | 0.0422*** (0.0090) | 0.0421*** (0.0093) | 0.0446*** (0.0052) | 0.0257*** (0.0021) | 0.0325*** (0.0161) |
| Hauschildt-Schlaak | 0.0012 (0.8760) | -0.0044 (0.5820) | -0.0046 (0.5640) | -0.0028 (0.7220) | -0.0056 (0.2400) | 0.0013 (0.8500) |
| contract | - | -0.0322 (0.2070) | -0.0322 (0.2060) | -0.0426* (0.0951) | -0.0201 (0.1630) | -0.0154 (0.4860) |
| trust | - | - | -0.0057 (0.8180) | 0.0003 (0.9910) | -0.0077 (0.5800) | -0.0089 (0.6700) |
| country | - | - | - | -0.121*** (0.0137) | -0.0136 (0.6310) | -0.136*** (0.0011) |
| constant | -0.402*** (0.0000) | -0.332*** (0.0026) | -0.329*** (0.0031) | -0.290*** (0.0082) | -0.181** (0.0104) | -0.277*** (0.0038) |
| n | 284 | 257 | 257 | 257 | 257 | 257 |
| LL | -139.86 | -123.58 | -123.56 | -120.59 | -64.32 | -101.79 |
| p | 0.0165 | 0.0252 | 0.0520 | 0.0084 | 0.0145 | 0.0051 |
| $R^2$ | 0.0271 | 0.0348 | 0.0350 | 0.0582 | 0.0846 | 0.0696 |
| Logit: Marginal effects for all SMEs with a successful innovation and cooperation partners. Probability *p* in brackets. | | | | | | |
| ***p<0.01, **p<0.05, *p<0.1 | | | | | | |

**Table 4:** Enterprises with less successful innovation probability of utilzation of "strong" vs "weak" ties

| predictors | (1) basic model | (2) + contract | (3) +trust | (4) + country | (5) + public consult | (6) + private consult |
|---|---|---|---|---|---|---|
| size | 0.0671$^{***}$ (0.0007) | 0.0616$^{***}$ (0.0022) | 0.0610$^{***}$ (0.0026) | 0.0608$^{***}$ (0.0025) | 0.0297$^{***}$ (0.0163) | 0.0410$^{***}$ (0.0088) |
| Hauschildt-Schlaak | 0.0062 (0.4750) | 0.0071 (0.4280) | 0.0069 (0.4400) | 0.0079 (0.3790) | 0.0096* (0.0774) | 0.0030 (0.6810) |
| contract | - | -0.0174 (0.5880) | -0.0175 (0.5840) | -0.0330 (0.3260) | -0.0023 (0.9190) | -0.0147 (0.5860) |
| trust | - | - | 0.0080 (0.8030) | -0.0030 (0.9250) | 0.0214 (0.3140) | -0.0343 (0.1710) |
| country | - | - | - | -0.0928 (0.1720) | -0.0104 (0.8130) | -0.1160$^{**}$ (0.0357) |
| constant | -0.541$^{***}$ (0.0000) | -0.534$^{***}$ (0.0000) | -0.530$^{***}$ (0.0000) | -0.498$^{***}$ (0.0000) | -0.392$^{***}$ (0.0000) | -0.345$^{***}$ (0.0007) |
| observations | 174 | 161 | 161 | 161 | 161 | 161 |
| LL | -85.43 | -77.54 | -77.51 | -76.58 | -47.70 | -60.47 |
| *p* | 0.0023 | 0.0104 | 0.0235 | 0.0236 | 0.0752 | 0.0085 |
| $R^2$ | 0.0645 | 0.0657 | 0.0660 | 0.0773 | 0.0846 | 0.1081 |
| Logit: Marginal effects for all SMEs with less successful innovation and cooperation partners. Probability *p* in brackets | | | | | | |
| $^{***}p<0.01$, $^{**}p<0.05$, $^{*}p<0.1$ | | | | | | |

## 5 Discussion

In general SMEs in an innovation process use by far rather the strong ties to customers and suppliers to initiate and enforce innovations than the weak ties to the consulting system. From the perspective of resource-oriented strategic management this cooperation behavior in the innovation process is coherent, as knowledge of potential or concrete innovations might diffuse via the weak ties and possibly drift to competitors. The study we present here also shows this decision behavior empirically: Both Danish and German SMEs utilize the strong ties much more than the weak ties when choosing the cooperation partners in the innovation process. In order to improve the utilization of the consulting system, a deeper understanding of the SMEs' cooperation behavior is essential. Here we argue with reference to the RDA that organizations will generally try to strengthen their external relations to other actors to avoid power dependencies and the influence associated with that. As a result, SMEs will only use the weak ties of the consulting system if they can control them or if they see a chance of evading power dependencies by using the consulting system. Based on the data that were used, it is almost exclusively the first case that can be observed empirically: generally SMEs will only build relations to the consulting system if they have strong cooperation relations at the same time. In contrast, it is only in exceptional cases that relations to the consulting system are recorded if there are no strong cooperation relations at the same time.

Based on the RDA a number of arguments were developed to provide a better explanation for the strategic management of the relationship to intermediaries and other cooperations partners in the process of innovation. The first assumption is that the supervision of external relations depends on the resource equipment of the organization, i.e. larger organizations should rather see themselves as being able to enter weak relations than comparatively smaller businesses. While this hypothesis is well confirmed, the more specific hypotheses are not confirmed in the same way. The argumentation that those SMEs that cannot secure their strong cooperation relations with formal (test criteria or contracts) or informal (mutual trust) control mechanisms will rather enter weak ties is not supported by the data analyses presented here. It is rather the mere presence of strong cooperation relations that will suffice to enter also weak relations. Neither is our further argumentation that the novelty and the uncertainty of the innovation process that is linked to it influence the cooperation behavior confirmed by the multivariate analysis. Comparing Denmark and Germany, however, the results of the multiva-riate analysis show that Danish SMEs utilize the consulting system, especially private consultancies, comparatively more often than German SMEs.

# 6 Implications

Practically these findings imply that the consulting system has difficulties in reaching smaller SMEs. This means that a considerable effort is required from public consultancies in particular to support innovations in SMEs. Based on this study it could not be clarified to which extent the decision behavior of SMEs indicates how the consulting system might be improved in other ways. This implies a need for research, as the conditions under which SMEs would wish for and would utilize consulting need to be clarified. To answer these questions a more differentiated argumentation might be necessary which also deals directly with the relations between SMEs and consultancies, not only indirectly with the cooperation relations with other partners. This argumentation was tailored to the research strategy of secondary analysis that was pursued here and which also accounts for part of the limits of this study. Certainly, the response to more profound questions requires another, extended database which provides more information about the behavior of SMEs in the innovation process and the utilization of the consulting system.

# References


Afuah, A. N. and Bahram, N, (1993) "The hypercube of innovation", Research policy, Vol. 24, pp 51-76.

Aldrich, H. E. and Pfeffer, J. (1976) "Environments of organizations", Annual review of sociology, Vol. 2, pp 79–105.

Brockhoff, K. (2003) "Customer´s perspectives of involvement in new product development", International Journal of Technology Management, Vol. 26, No. 5/6, pp 464–481.

Coleman, J. S. (1974). Power and Structure of Society. New York: Norton.

DiMaggio, P. J. and Powell, W. W. "The iron cage revisited: Institutional isomorphism and collective rationality in organizational fields", American Sociological Review, Vol. 48, pp 147-160.

Duschek, S. (2004). "Inter-firm resources and sustained competitive advantage". Management Revue, Vol. 15, pp 53–73.

Dyer, H. J. and Singh, H. (1998) "The relational view: Cooperative strategy and sources of interorganizational competitive advantage", Academy of Management Review, Vol. 23, Nr. 4, pp 660–679.

Emerson, R. M. (1962) "Power dependence relations", American Sociological Review, Vol. 27, pp. 31–40.

Freiling, J. (2008) "RBV and the road to the control of external organizations" Management Revue, Vol. 19 Nr. 1/2, 33–52.

Glückler, J. and Armbrüster, T. (2003) "Bridging uncertainty in management consulting: The mechanisms of trust and network reputation", Organization Studies, Vol. 24, Nr. 2, pp 269-297.

Hannan, M. T., Freeman, J. (1984). "Structural Inertia and Organizational Change", American Sociological Review, Vol. 49, No. 2, pp 149–164.

Hauschildt, J. and Schlaak, T. M. (2001) "Zur Messung des Innovationsgrades neuartiger Produkte", Zeitschrift für Betriebswirtschaft, Vol. 71 No. 2, pp 161–182.

Hippel, v. E. (1978) "A customer active paradigm for industrial product idea generation. In: Research Policy, 7, 240–266.

Hislop, D. (2002). "The client role in consultancy relations during the appropriation of technological innovations", Research Policy, Vol. 31, pp 657–671.

Nienhüser, W. (2008) "Resource dependence theory: How well does it explain behavior of organizations?" management revue, Vol. 19, pp 9–32.

Pfeffer, J. and Salancik, G. R. (1978) The External Control of Organizations. A Resource Dependence Perspec-tive, New York.

Reichwald, R. and Piller, F. (2005) "Open Innovation: Kunden als Partner im Innovationsprozess" Working paper, DGF-Sonderforschungsbereich 582, pp 1-19.

Poulfelt F, Payne A (1994), "Management consultants: Client and consultant perspectives", Scandinavian Journal of Management, Vol. 10, No. 4, pp 421–436.

Schumpeter, J. A. (2006) Theorie der wirtschaftlichen Entwicklung, Berlin.

Wernerfelt, B. (1984) "A resource-based view of the firm" Strategic Management Journal, Vol. 5, No. 2, pp 171–180.

Williamson, O. E. (1985) The Economic Institutions of Capitalism: Firms, Markets, Relational Contracting. New York.

Wolpert, J. P. (2002), "Breaking out of the innovation box" Harvard Business Review, Vol. 80, Nr. 8, pp 76–83.

Wu, L. and Lin C.-Y. and Aral, S. and Brynjolfsson, (2009) "Value of Social Network – A large-scale analysis on network structure impact to financial revenue of information consultants", Conference paper, presented at the Winter Information Systems Conference, Salt Lake City, UT/US.

# Knowledge Management and Open Innovation in a Bioengineering Research Case

**Manel González-Piñero, Elena López Cano, Miguel Ángel Mañanas Villanueva, Juan Ramos Castro and Pere Caminal Magrans**
Technical University of Catalonia, Barcelona, Spain

**Editorial commentary**

In this detailed case study, the authors explore, and reflect upon, the interactions taking place between the various players in the research, development, intellectual property, and commercialisation processes of a technology (Biofeedback equipment) for measuring urinary incontinence. In particular, the study focuses on the role played by a university in the knowledge transfer process: from academia, through industry (manufacture) and on to market, and highlights the importance of ensuring good communication and good relationships between key stakeholders throughout the R&D and commercialisation processes.

Some points for discussion and learning arising from this case study include:

- The contribution of each stakeholder in a successful commercialisation project involving universities and industry.
- The role of knowledge management in such a commercialisation project.
- The factors that lead to successful knowledge transfer between universities and industry.
- The barriers to knowledge transfer in a collaborative project between a university and key industrial partners.

**Abstract**: One of the missions of University is the transfer of technologies to the industry in order to be profitable for the society, in terms of economy, productivity and social advances. The Bioengineering case study that we expose has completed all the steps of the Valorisation chain: detection, selection, evaluation, protection and commercialization. In this process we show how the interaction between researchers, IP managers and business development managers is significant to know the market requirements and to direct the research efforts towards the market necessities; that is why the last step of the design is a commercialization planning to explode the technology. In this case, the goal is to find an industrialized partner interested in producing the clinical validated prototype following all the EC regulations. The knowledge selected to advance in the value chain before being transferred to the market is a technology of a Biofeedback equipment for urinary incontinence. These devices for the treatment of urinary incontinence have the purpose of generating (and also recording for further analysis) signals directly related to the activity of the pelvic floor muscles. Although a few instruments use the exerted pressure as the measured signal, the most frequently employed signal is the electrical activity associated to the muscle contractions or electromyogram (EMG).
**Keywords**: bioengineering, biofeedback, entrepreneurship, marketing in higher education, technology transfer, knowledge transfer, valorisation process

# 1 Introduction

Universities are an agent at the Triple Helix system, a model designed by Etzkowitz, (2002) and Leydesdorff (2003), in which the knowledge management activities are basics and have to be integrated into the business processes, due to the fact that the knowledge created in the University has the ambition to be profitable at the society. This theory about innovation systems is based on university-industry-government relations, and the generation of knowledge and its effective application. So, how it can be possible the technology transfer of the technological opportunities identified in the university? It could be possible through an effective commercial model defined in a Valorisation Process.

The knowledge selected to advance in the value chain before being transferred to the market is a technology of a Biofeedback equipment specifically for urinary incontinence. It is placed in a female's vagina or the anus of a male and connected to a computer or a portable home therapy unit to give the person a light signal, bar graph, or audible tone to show how the muscles of the pelvic floor are contracting and relaxing. Surface electrodes can also be placed on the abdomen or the buttocks and used in conjunction with internal sensors. The instrument accurately measures, processes and 'feedback' to the patient and clinician the data in the form of visual and/or auditory signals. This information has educational and reinforcing characteristics about the muscle activity. The knowledge of muscle activity allows the patient and clinician to adjust existing motor patterns by re-educating the muscle or muscle groups for efficient muscle function and voluntary control.

Biofeedback instruments for the treatment of urinary incontinence have the purpose of generating (and also recording for further analysis) signals directly related to the activity of the pelvic floor muscles. Although a few instruments use the exerted pressure as the measured signal, the most frequently employed signal is the electrical activity associated to the muscle contractions or electromyogram (EMG).

The study wants to put forward how a real biomedical technology developed at the Biomedical Engineering Research Centre of the Technical University of Catalonia is finally successfully transferred to the market, showing how researchers, IP managers and business development managers work together to achieve a common aim.

## 2 The value chain

This concept from business management was first described and popularized by Porter (1985: 36) “Every firm is a collection of activities that are performed to design, produce, market, deliver, and support its product All of these activities can be represented using a value chain...”. Porter termed the larger interconnected system of value chains the "value system" (1985: 34) when states that “A firm’s value chain is embedded in a larger stream of activities that I term the value system...”. A value system includes the value chains of a firm's supplier (and their suppliers all the way back), the firm itself, the firm distribution channels, and the firm's buyers (and presumably extends to the buyers of their products, and so on).

Value Streams were introduced in Porter’s work but were explained more clearly by Martin (1995: 66), which pulls together the many issues, models, and methods for transforming the traditional old-world organization into a value-creating enterprise. Martin uses value stream, rather than process, to define the end-to-end stream of activities that deliver particular results for a given customer (external or internal).

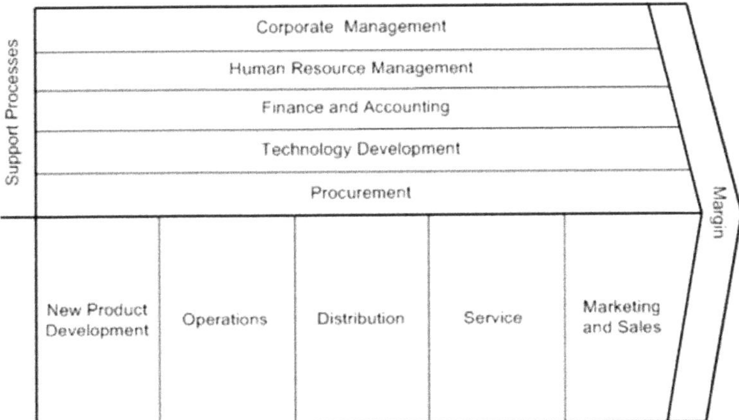


**Figure 1:** Porter's value chain

Value chain is a high-level model of how businesses receive raw materials as input, add value to the raw materials through various processes, and sell finished products to customers. The value chain categorizes the generic value-adding activities of an organization.

A value chain is the disaggregating of a firm into its strategically relevant activities for the purpose of understanding the behaviour of costs as well as the existing and potential sources of differentiation. The concept has been extended beyond individual organizations. The industry wide synchronized interactions of those local value chains create an extended value chain, sometimes global in extent. Capturing the value generated along the chain is the new approach taken by many management strategists. By exploiting the upstream and downstream information flowing along the value chain, the firms may try to bypass the intermediaries creating new business models.

The chain of the Technology Valorisation Process contains the following five steps: detection, selection, evaluation, protection and commercialization.

## 3 Detecting a problem: A R&D and business opportunity

Urinary incontinence is defined by the International Continence Society (ICS) as a 'condition in which involuntary loss of urine is a social or hygienic problem and is objectively demonstrable' (Abrams et al. 1988). It is often a devastating condition severely affecting the quality of life for sufferers.

The most common form of urinary incontinence (affecting approximately 45% of incontinent people), is stress incontinence, described as a leakage of urine associated with an increase in intra-abdominal pressure for example during a cough sneeze, laugh or physical exertion. If the client is investigated using subtracted cystometry (urodynamics) where the behaviour of the Detrusor muscle can be observed, then a diagnosis of genuine stress incontinence (GSI), may be made. This is in accordance with the ICS definition of genuine stress incontinence (Abrams, 1988): “GSI is the involuntary loss of urine occurring when in the absence of detrusor contraction, the intravesical pressure exceeds the intraurethal pressure.” Acknowledging the role of the change in intra-abdominal pressure, (Abrams, 1999) suggests that this definition should be modified to read: ¨The involuntary loss of urine occurring when due to raised intra-abdominal pressure, the intravesical pressure exceeds the intraurethral pressure.¨

Urinary incontinence is a common problem, estimated to affect more than 13 million people in the United States - male and female, young and old - and a condition more prevalent than diabetes mellitus (National Kidney and Urologic Diseases Information Clearinghouse, 2002). Overactive bladder alone, which includes patients who do not leak urine, is reported to affect 17 million Americans. Because of the recognized phenomenon of underreporting, these numbers may be low. Millions more people are not incontinent but would be if they did not continually void to prevent leakage. This reality is mirrored in countries worldwide (see Fig. 2 from Hampel et al, 1999). Patients with this problem often lead lives of quiet desperation and social isolation.

The National Association for Continence (NAFC) states “that while only one out of twelve incontinent patients in the United States actually reports their symptoms to their doctors, approximately 80% can be cured or improved”. Unfortunately, women wait an average of 3 years before admitting their incontinence to a health care provider (Woolner et al, 2001). A persistent myth is that incontinence is a natural part of aging.

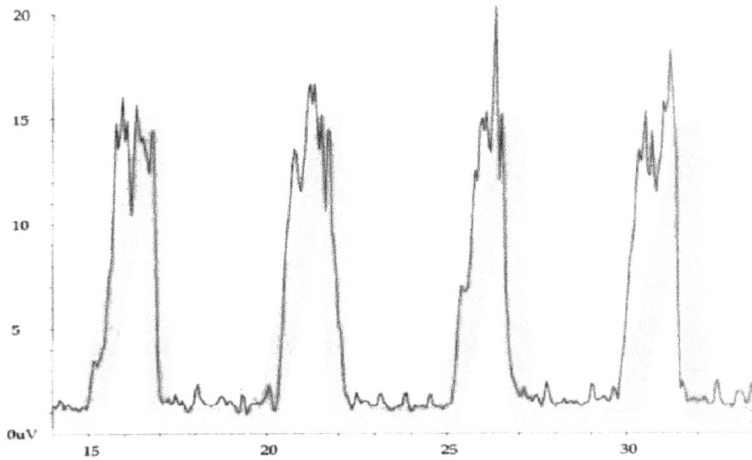


**Figure 2:** Example of template training on the screen

Urinary incontinence not only causes considerable medical and psychosocial morbidity but it also engenders enormous costs (Resnick, 1998) (Urinary Incontinence Guideline Panel 1992). In 1994, it was estimated that $11.2 million was spent on the direct treatment of incontinence, and $5.2 million on associated nursing home costs. Wagner and Hu at the University of California at Berkeley documented the direct cost of incontinence in the Medicare-age population to be more than $16 billion per year [Wagner and Hu, 1998]. When indirect costs (eg, time lost from work) were added, the figure rose to more than $26 billion per year, and that only included senior citizens. It is more than is expended by Medicare on dialysis, and coronary bypass grafting combined (Health Care Financing Review, 1997). Moreover, these costs apply only to individuals older than 65 years, who constitute less than half of those with urinary incontinence.

For older adults living in the community, the prevalence of urinary incontinence is between 15% and 35%, with women affected twice as often as men (Lefevre, 2000). The condition is even more common among residents of nursing homes, where more than half of the residents experience urinary incontinence. The prevalence of urinary Incontinence depends on its definition (Hampel et al., 1999) (see Fig. 3). In addition, urinary incontinence has been cited as one of the major precipitants for placement in a nursing home (Ouslander et al. 1982). Thus, among the elderly Medicare population, this condition is associated with a high burden of illness, high costs, and has a substantial effect on quality of life.

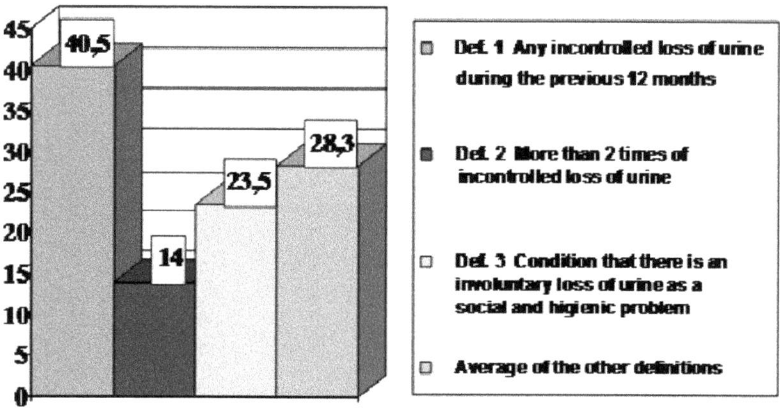


**Figure 3:** Prevalence of urinary incontinence (from [Hampel et al., 2009])

Urinary incontinence (UI) affects up to one-third of the non institutionalized population over age 60, with 25 percent to 30 percent of these individuals having frequent urinary incontinence episodes (US Dept. of Health and Human Services, 1996). UI may vary from non-significant to severe, causing extreme activity limitation and social isolation.

Because urinary incontinence impacts function, it is a primary predictor of decline in elderly people (Tinetti et al, 1995). Beyond the patient, urinary incontinence may also cause significant psychosocial distress to family, friends, and caregivers. National data show the annual direct cost of UI exceeds $11 billion dollars for community dwelling persons alone (US Dept. of Health and Human Services, 1996).

While it is estimated that the number of incontinent geriatric patients can be as high as 80% (Portnoi, 1981), it is more difficult to estimate the incidence in younger populations, though studies by Nygaard show incontinence to be common in young nulliparous women, particularly during physical activities. One Danish study (Hording, 1986), conducted with a group of 45-year-old women, found that 22% experienced stress incontinence. It was also noted that only three percent of these women sought medical attention for their problem.

Unfortunately, most physicians have received little education about incontinence, fail to screen for it, and view the likelihood of successful treatment as low (Fantl et al, 1996). Thus, it is fitting that the study by Burgio et al, which examines the efficacy of biofeedback for urge incontinence, should appear in (Burgio et al, 1998).

Incontinence is experienced more often when women are older. But incontinence is not inevitable with age. Incontinence is treatable and often curable at all ages. An understanding of the physiologic forces that leave the patient predisposed to incontinence is necessary for proper treatment selection.

# 4 Specifications of a biofeedback for urinary incontinence

Biofeedback equipment specifically for urinary incontinence requires a sensor that is placed in a female's vagina or the anus of a male and connected to a computer or a portable home therapy unit to give the person a light signal, bar graph, or audible tone to show how the muscles of the pelvic floor are contracting and relaxing. Surface electrodes can also be placed on the abdomen or the buttocks and used in conjunction with internal sensors. In this instance, the surface electrodes help the person eliminate the use of muscles that do not help with bladder control, and the internal sensors help focus the person's attention on contracting the correct muscles of the pelvic floor. The instrument accurately measures, processes and 'feedback' to the patient and clinician the data in the form of visual and/or auditory signals. This information has educational and reinforcing characteristics about the muscle activity. The knowledge of muscle activity allows the patient and clinician to adjust existing motor patterns by reeducating the muscle or muscle groups for efficient muscle function and voluntary control. The sEMG noninvasively monitors aggregated muscle activity for the purpose of functional evaluations and rehabilitation.

Biofeedback instruments for the treatment of urinary incontinence have the purpose of generating (and also recording for further analysis) signals directly related to the activity of the pelvic floor muscles. Although a few instruments use the exerted pressure as the measured signal, the most frequently employed signal is the electrical activity associated to the muscle contractions or electromyogram (EMG).

The electrical signals from skeletal muscle occur at frequencies audible to humans and can be amplified and played through a loudspeaker. This is the known as audible raw EMG signal employed in some systems as an output of the biofeedback system. Sounds associated with abnormal muscle or nerve tissue are often more distinctive and characteristic than visual signals and can be of more value than displayed waveforms in identifying specific disorders; the audio and visual input also helps the operator place the electrode.

# 5 Benefits of biofeedback vs. medication

There are multiple behavioural techniques and protocols, but their comparative efficacy is unknown. Because technical aspects cannot be detailed sufficiently in reports of clinical trials, behavioural interventions also are difficult to replicate in practice. One behavioural technique, biofeedback, has been even less widely used for urge incontinence because it often has required repeated instrumentation of the bladder and urinary sphincter (Resnick, 1998). Moreover, despite the expertise and time entailed, behavioural techniques are poorly reimbursed. By contrast, pharmacotherapy works more quickly and also requires no behavioural expertise, less physician time, and less patient participation. Nonetheless, although drugs help most patients, non drug restores continence to the majority. Furthermore, all of the agents currently used engender adverse effects, expense, and inconvenience (Fantl et al., 1996), and most must be taken several times daily and indefinitely. Thus, an equally or more effective one-time intervention would be welcome.

The study by Burgio (Burgio et al., 1998) demonstrates that a less-invasive biofeedback approach can achieve these goals. The study also serves as a rich source of information and provides valuable lessons for clinicians. The investigators' decision to begin therapy with oxybutynin at 2.5mg3 times daily was wise. The effect of oxybutynin was equivalent to that achieved in trials using higher dosages, but it caused far fewer adverse effects and a lower rate of subject attrition (Fantl et al. 1996). Equally important, efficacy continued to increase beyond 2 weeks, longer than previously reported, but consistent with other recent data (Appell, 1997) (Abrams et al., 1998). Thus, clinicians should avoid escalating dosages of oxybutynin too quickly or abandoning pharmacotherapy too soon.

# 6 The open innovation process followed in the R&D process of a biofeedback for urinary incontinence

"Open innovation is a paradigm that assumes that firms can and should use external ideas as well as internal ideas, and internal and external paths to market, as the firms look to advance their technology", as Chesbrough (2003: Introduction) defined in his work. The boundaries between a firm and its environment have become more permeable; innovations can easily transfer inward and outward. The central idea behind open innovation is that in a world of widely distributed knowledge, companies cannot afford to rely entirely on their own research but should instead buy or license processes or inventions (i.e. patents) from other companies.

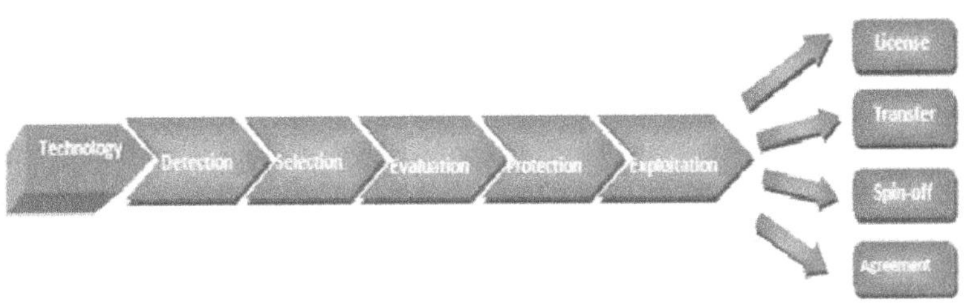


**Figure 4:** Technical University of Catalonia's value chain

**A) Detection:**

In a first phase, we would need to know which technologies are mostly thought out, or in a developed state of research. Those developed technologies would be in optimal conditions to be transferred to the market, because they have been got over the theoretical phase to arrive to the industrial implementation.

University has a huge technology stock most of the time not profitable for the market. Traditionally, the technology transfer of the university to the market has been concentrated basically in collaborative agreements with industries to solve ad hoc necessities.

Nowadays in our universities remain a lot of developed technologies ready to be exploited commercially. Our university has lately detected a large number of technologies thanks to the new grants offered by our regional Government of Catalonia, which goal is the valorization of technologies created in the core of research centres, like universities, hospitals or technology centres in order to transfer them to the Catalan industry to make them internationally more competitive (González-Piñero, 2009).

A first development of the biofeedback was developed under the petition of a Catalan SME that detected the market opportunities of this new medical device. Two divisions of the Biomedical Engineering Research Centre (CREB) of the Technical University of Catalonia (the Biomedical Signals and Systems Division and the Instrumentation and Bioengineering Division) are the leaders of this project in which also the Hospital Clínic de Barcelona would be part of the project as a responsible of the validation process at the hospital.

**B) Selection**

We must consider that not all the technologies satisfy the optimal commercial parameters to be exploited. That is the reason to evaluate the technology detected and to do a "Proof of concept" if we want to continue in the value chain. This evaluation consist in two aspects: one related with marketing and science-business, and another that evaluate the technology quality. The first evaluation is done for experts in marketing and business in the university, and the second evaluation for a Scientific Committee.

In our case, we have studied all the possibilities that currently are implanted in the market. It will be helpful to define our strategy to achieve an innovative methodology and final prototype.

There are a lot of devices in the market for biofeedback treatment; the most part of them are from USA, where there exists a great demand. We have included the devices than can be applied, as specified by the manufacturer, in the treatment of urinary incontinence.

The instruments have been divided into three categories according to their use.

- Computer based: They are not intended for home use. Its use is restricted to practitioner consult. Normally they are bulk devices. They can record different signals (EMG, pressure, temperature, skin conductivity, etc.) and the user interface is build around a PC. Table 17 summarizes the main characteristics for several manufacturers. Prices range from 3000 $ to 8000 $ and their use is not oriented specifically for urinary incontinence treatment.
- Portable: They are intended for home use and their main characteristics are easy operation, simple user interface (normally a bar led and 1 or 2 buttons) 1 or 2 EMG channels. Table 18 summarizes the main characteristics for several manufacturers. Due to the differences in specifications between manufacturers prices range from 300 $ to 1300 $.
- Portable with computer interface: They are intended for home and ambulatory use and their main characteristics are easy operation, user graphical interface (normally a bar led and a LCD display) 2 EMG channels and an interface link to connect to a PC where a specific software can be used for monitoring online, download recorded data or program different protocols. Table 19 summarizes the main characteristics for several manufacturers.

After this process, the final specifications of a biofeedback system for urinary incontinence have been defined and we have started the R&D process under this assumption. The validation stage has been developed at the Hospital Clinic, under the supervision of doctors and the researchers of the CREB.

The project is funded by the Catalan Government to foster transfer technology strategies in research applied projects.

**C) Evaluation**

- This business case of this new technology requires to pay attention on three areas that will define the impact of the technology once launched into the market:
- Industrial Property: Analysis of the state-of-the-art, other existing patents concerning the technology, family patents.
- At the moment, we are studying with a consulting firm the patentability of the final device. It has many possibilities and the results of the R&D can be protected and there is the freedom to operate. A national patent has been requested.
- Market prospection: interesting sectors, advantages of our technology respect another solutions in the market, industrial applications, time-to-market, competitors, risks, potential clients.
- This stage has been developed by a consulting firm (KIMBCN) in order to detect which companies could be interested in the acquisition of this device. This market study has been opened to firms around the world and some initial contacts have been established with some of them.
- Legal Aspects: rules and regulations. The final prototype has to be validated and follow the EU regulations.

**D) Protection**

Considering the protection aspects of the technology is a key aspect. Firstly, we must know if the technology is developed in collaboration with other partners in order to negotiate the co-property percentages. Secondly, it is necessary to consider the different ways of protection (patent, design, software ...), or to take into account the possibility of non protection (industrial secret). Finally, we have to evaluate if the technology fulfils all the requirements to be patented (invention must be novel, invention must not be obvious to someone with a good knowledge of the subject, invention must have an industrial use).

In our case, we are considering the possibility of licensing the patent. Licensing occurs when a licensor grants exploitation rights over a patent to a licensee. This is also a legal contract, and so it will set out the terms upon which the exploitation rights are granted, including performance obligations that a licensee must comply with. Therefore, the failure to comply with those obligations may lead to the termination of the license, and the reversion of exploitation rights back to the licensor.

**E) Exploitation**

The final step in the chain of value is to consider the exploitation of the technology. Not all the technologies can be exploited in the same way, we have different formulas:

- License (in all its forms)
- Transfer to a customer (sale) keeping a free license use
- Spin-off
- Collaborative agreements with companies to industrialize a final product. The industry participates in the final development in order to exploit later together.

In our case, we want to transfer the technology to a company well positioned at the market and with a good knowledge of the specifications of the product, demands of the market and business opportunities.

# 7 Regulations and certifications

The European Directive 93/42/CEE of 14 June 1993 concerning medical devices makes it mandatory to fulfill CE certification requirements in order to product medical devices, of Class I, IIa, IIb, and III to export them to any country within the European community.

The Medical Device Directive, MDD 93/42/EEC, becomes mandatory on March 21, 2010. This means that all medical device manufacturers have to review their compliance to this directive and be in compliance. Otherwise they have to stop selling their devices in the European community. This is especially critical for the new classification rules, if a Class I device is now classified as Class II a or b, or from Class II to Class III, a new conformity assessment is required.

According to Directive 2007/47/EC which has amended the Directive 93/42/EEC, Medical Device means: any instrument, apparatus, appliance, software, material or other article, whether used alone or in combination, including the software intended by its manufacturer to be used specifically for diagnostic and/or therapeutic purposes and necessary for its proper application, intended by the manufacturer to be used for human beings for the purpose of:

- Diagnosis, prevention, monitoring, treatment or alleviation of disease,
- Diagnosis, monitoring, treatment, alleviation of or compensation for an injury or handicap,
- Investigation, replacement or modification of the anatomy or of a physiological process,
- Control of conception,

and which does not achieve its principal intended action in or on the human body by pharmacological, immunological or metabolic means, but which may be assisted by such means.

Medical Devices shall be divided into Classes I, IIa, IIb and III (see 93/42/EEC), which shall be carried out in accordance with Annex IX. Examples of devices in each type:

- Class I: Wheel Chairs Patient electrodes. Scalpels. Dental Drills. Wound Management systems. Hearing Aid Tester.
- Class IIa: All patient monitoring equipment, Syringes. Needles. Ultra Sound devices. External ECGs. Diagnosis devices mainly.
- Class IIb: Lasers Devices for application. Internal ECGs. RF Generators. Non-energized implants. Treatment devices mainly.
- Class III: Energized implants All Intracardiac applications. Heart valves. Cauterters. Non-energized implants. All devices in contact with the central nervous system.

Member States shall take all necessary steps to ensure that devices may be placed on the market and put into service only if they do not compromise the safety and health of patients, users and, where applicable, other persons when properly installed, maintained and used in accordance with their intended purpose, following the requirements set out in the Annexes of the Directive 93/42/CEE.

# 8 Conclusions

The technologies developed at the University must follow a methodology of Valorization if they want to be successfully exploited. That is why a good interaction between all the actors involved in the process like researchers, IP managers and business development managers is crucial. It is also significant to know the requirements of the market to direct the research efforts towards the market necessities because demands from industry, hospital and final users are the key to be successful in applied research and the subsequent transfer.

Currently, the Biofeedback is under prototyping and clinical trial has been successful. At the moment, the process is being finished to apply the National patent and a final planning of commercial activities has been developed. Now, we are in contact with industry partners to negotiate the production of the prototype and the license of the patent. But at the same time we are studing the possibility to create a spin-off to commercialize this technology and others developed by ourselves related with other biomedical devices.

# References

Abrams P, et al. (1988:114: 5) The Standardisation of terminology of lower urinary tract function. Scand J Urol Nephrol (Suppl).

Burgio et al (1998, 280(23):1995-2000) Behavioral vs Drug Treatment for Urge Urinary Incontinence in Older Women: A Randomized Controlled Trial, The Journal of the American Medical Association.

Chesbrough, H. (2006) Open Innovation: Researching a New Paradigm, Oxford.

Chesbrough, H. (2003) Open Innovation: The New Imperative for Creating and Profiting from Technology Harvard Business School Press.

COUNCIL DIRECTIVE 93/42/EEC of 14 June 1993, concerning to medical devices.

Council and Parliament Directive 2007/47/EC of 5 September 2007, amending Council Directive 90/385/EEC on the approximation of the laws of the Member States relating to active implantable medical devices, Council Directive 93/42/EEC concerning medical devices and Directive 98/8/EC concerning the placing of biocidal products on the market.

Etzkowitz, H. (2002) The Triple Helix of University – Industry – Government Implications for Policy and Evaluation ISSN 1650-3821.

Fantl et al, (1996) Urinary Incontinence in Adults: Acute and Chronic Management, Clinical Practice Guideline nº 2.

Geiger, R.(2004) Knowledge and Money: Research Universities and the Paradox of the Marketplace, Stanford University Press.

González de la Fe, T. (2009) Triple Helix Model of relations among University, Industry and Government: a critical Analysis.

González-Piñero, M. (2009) Tecnologies Mèdiques. Una oportunitat estratègica per a l'empresa catalana. III Congrés d'Enginyeria i Cultura Catalana.

Hampel, C. et al (2009) Population-Based Survey of Urinary Incontinence, Overactive Bladder, and Other Lower Urinary Tract Symptoms in Five Countries: Results of the EPIC Study, European Urology, Volume 50, Issue 6

Hampel, C. et al, (1999:10–15) Heterogeneity in epidemiological interventions of bladder control problems: a problem of definition. BJU Int 83 Suppl 2

Hörding U, et al (1986:183–186), Urinary incontinence in 45-year-old women, Scand J Urol Nephrol 20

Martin J. (1995) The Great Transition: Using the Seven Disciplines of Enterprise Engineering to Align People, Technology, and Strategy.

Lefevre F. (2000) Biofeedback in the treatment of urinary incontinence in adults, Blue Cross and Blue Shield Association.

Leydesdorff, L A, (2003:201-204) Methodological Perspective on the Evaluation of the Promotion of University-Industry-Government Relations.Small Business Economics 20 (2),.

Ouslander et al. (1982), Urinary incontinence in elderly nursing home patients, The Journal of the American Medical Association.

Porter M. (1985:64- 66) Competitive Advantage: Creating and Sustaining Superior Performance.

Portnoi, V.A., (1981: 23:151-154) Urinary incontinence in the elderly. Am. Fam. Physician.

Resnick NM. (1998; 280:2034 5) Improving treatment of urinary incontinence. The Journal of the American Medical Association.

Rubiralta M, (2007:27-41) La transferencia de la I+D en España, principal reto para la innovación, Economía Industrial, nº 366.

Rubiralta M et al. (2003), Nuevos mecanismos de transferencia de tecnología: Debilidades y oportunidades del sistema español de transferencia de tecnología. Madrid, COTEC.

Solé Parellada, et al. (2006:66-83) Valorització de la recerca a la universitat: és útil la recerca més enllà de les publicacions que fan els investigadors? Coneixement i Societat: Revista d'Universitats, Recerca i Societat de la Informació, nº 12,

Tinetti, M. E., et al. (1995:1348–1353) Shared risk factors for falls, incontinence, and functional dependence unifying the approach to geriatrics syndromes. Journal of the American Medical Association, 273.

Wagner TH, et al (1998;51:355–61) Economic costs of urinary incontinence in 1995. Health Services and Policy Analysis Program, School of Public Health, University of California at Berkeley.

www.ingramcontent.com/pod-product-compliance
Ingram Content Group UK Ltd.
Pitfield, Milton Keynes, MK11 3LW, UK
UKHW020130250726
13967UKWH00002B/562